An Introduction to the Formalism of Quantum Information with Continuous Variables

An Introduction to the Formalism of Quantum Information with Continuous Variables

Carlos Navarrete-Benlloch

Max-Planck Institute of Quantum Optics, Garching by Munich

Morgan & Claypool Publishers

Rights & Permissions
To obtain permission to re-use copyrighted material from Morgan & Claypool Publishers, please contact info@morganclaypool.com.

ISBN   978-1-6817-4405-6 (ebook)
ISBN   978-1-6817-4404-9 (print)
ISBN   978-1-6817-4407-0 (mobi)

DOI   10.1088/978-1-6817-4405-6

Version: 20151201

IOP Concise Physics
ISSN 2053-2571 (online)
ISSN 2054-7307 (print)

A Morgan & Claypool publication as part of IOP Concise Physics
Published by Morgan & Claypool Publishers, 40 Oak Drive, San Rafael, CA, 94903, USA

IOP Publishing, Temple Circus, Temple Way, Bristol BS1 6HG, UK

To all my colleagues, friends, and family who made this book possible.

Contents

Preface   ix

Acknowledgements   xi

Author's biography   xiii

1   Quantum-mechanical description of physical systems   **1-1**

1.1   Classical mechanics   1-1

     1.1.1   The Lagrangian formalism   1-1

     1.1.2   The Hamiltonian formalism   1-3

     1.1.3   Observables and their mathematical structure   1-4

1.2   The mathematical language of quantum mechanics   1-6

     1.2.1   Finite-dimensional Hilbert spaces   1-6

     1.2.2   Linear operators in finite dimensions   1-8

     1.2.3   Generalization to infinite dimensions   1-11

     1.2.4   Composite Hilbert spaces   1-13

1.3   The quantum-mechanical framework   1-14

     1.3.1   A brief historical introduction   1-14

     1.3.2   Axiom 1. Observables and measurement outcomes   1-15

     1.3.3   Axiom 2. The state of the system and statistics of measurements   1-16

     1.3.4   Axiom 3. Composite systems   1-18

     1.3.5   Axiom 4. Quantization rules   1-19

     1.3.6   Axiom 5. *Free* evolution of the system   1-20

     1.3.7   Axiom 6. Post-measurement state   1-22

     1.3.8   The von Neumann entropy   1-25

     Bibliography   1-27

2   Bipartite systems and entanglement   **2-1**

2.1   Entangled states   2-1

2.2   Characterizing and quantifying entanglement   2-3

2.3   Schmidt decomposition and purifications   2-5

     Bibliography   2-7

3   Quantum operations   3-1

3.1   Basic principles of quantum operations   3-1
    3.1.1 General considerations   3-1
    3.1.2 Further properties of quantum operations   3-3
    3.1.3 Quantum operations as reduced dynamics in an extended system 3-4
3.2   Generalized measurements and positive operator-valued measures   3-5
3.3   Local operations and classical communication protocols   3-6
3.4   Majorization in quantum mechanics   3-7
    3.4.1 The concept of majorization   3-8
    3.4.2 Majorization and ensemble decompositions of a state   3-8
    3.4.3 Majorization and the transformation of entangled states   3-9
    Bibliography   3-10

4   Quantum information with continuous variables   4-1

4.1   The classical harmonic oscillator   4-1
4.2   The quantum harmonic oscillator   4-2
4.3   The harmonic oscillator in phase space: the Wigner function   4-7
    4.3.1 General considerations   4-7
    4.3.2 Definition based on the characteristic function   4-9
    4.3.3 Multi-mode considerations   4-10
4.4   Gaussian continuous-variable systems   4-12
    4.4.1 Gaussian states   4-12
    4.4.2 Gaussian unitaries   4-17
    4.4.3 General Gaussian unitaries and states   4-23
    4.4.4 Gaussian bipartite states and Gaussian entanglement   4-27
    4.4.5 Gaussian channels   4-33
4.5   Measuring continuous-variable systems   4-38
    4.5.1 Description of measurements in phase space   4-38
    4.5.2 Photodetection: measuring the photon number   4-39
    4.5.3 Homodyne and heterodyne detection: measuring the   4-42
           quadratures and the annihilation operator
    4.5.4 On/off detection and de-Gaussification by vacuum removal   4-45
4.6   Non-Gaussian scenarios: photon addition, subtraction, and   4-48
    majorization properties of two-mode squeezed states
    4.6.1 Photon addition and subtraction   4-48
    4.6.2 Increasing entanglement by local addition or subtraction   4-50
    4.6.3 Majorization properties of two-mode squeezed Fock states   4-52
    Bibliography   4-55

Preface

Quantum information is an emerging field which has attracted a lot of attention in the last couple of decades. It is a broad subject which extends from the most applied questions (e.g. how to build quantum computers or secure cryptographic systems), to the most theoretical problems concerning the formalism and interpretation of quantum mechanics, its complexity, and its potential to go beyond classical physics.

This book is devoted to the introduction of the formalism behind quantum information, with special emphasis on continuous-variable systems, that is, systems (such as light) which can be described as collections of harmonic oscillators. It must not be taken as an exhaustive review of the field, but rather as an introductory text covering a selection of basic concepts, focusing on their physical meaning and mathematical treatment. The book is intended to be self-contained, in the sense that it starts from the very first principles of quantum mechanics, and tries to build up the concepts and techniques following a logical progression. Consequently, the target audience for this book is, primarily, students who have already studied a full semester of standard quantum mechanics and want to delve deeper into it, or researchers in closely related fields such as quantum optics or condensed matter, who would like to learn about the powerful formalism of quantum information. Length restrictions have precluded me from including practical examples that could clarify the definitions and statements, but these can always be found in the various references that I include in the different sections.

As for the choice of topics, I have chosen those that I found the most valuable when I first approached the field as a PhD student in quantum optics. The book starts by reviewing in the first chapter some basic notions of classical mechanics, linear algebra, and quantum mechanics that are needed throughout the text. Two chapters dealing with concepts applicable to general quantum systems follow, one related to bipartite entanglement, and the other to the formalism of general quantum operations (the most general transformations that one can induce on a system), including the important examples of channels and generalized measurements. The last chapter, which is the main chapter of the book, applies the ideas developed in the previous chapters to continuous-variable systems, that is, systems such as harmonic oscillators described by infinite-dimensional Hilbert spaces. After reviewing the physics of the quantum harmonic oscillator, a phase-space description based on the Wigner function is discussed. The relevant class of Gaussian states and operations is introduced next, emphasizing how they are very important both for theoretical and applied reasons, since they are relatively easy to treat mathematically and generate experimentally. After this, the quantum theory of measurements is particularized to continuous-variable systems, introducing typical detection strategies in scenarios such as photo-, homodyne-, heterodyne-, and on/off-detection. Finally it is shown that detection strategies can be used to generate non-Gaussian operations such as photon addition and subtraction probabilistically, which in turn can be used to enhance entanglement when applied locally, against intuition.

Thorough reviews covering further topics and techniques in the field of continuous variables, as well as some proofs that I have skipped, can be found in [1–6].

Above all, I hope interested readers will find this book an enjoyable ride through the thrilling subject of continuous-variable quantum information.

Carlos Navarrete-Benlloch
Munich, November 2015.

Bibliography

[1] Braunstein S L and van Loock P 2005 Quantum information with continuous variables *Rev. Mod. Phys.* **77** 513

[2] Weedbrook C, Pirandola S, García-Patrón R, Cerf N J, Ralph T C, Shapiro J H and Lloyd S 2012 Gaussian quantum information *Rev. Mod. Phys.* **84** 621

[3] Ferraro A, Olivares S and Paris M G A 2005 *Gaussian States in Quantum Information* (*Napoli Series on Physics and Astrophysics*) (Naples: Bibliopolis)

[4] Cerf N J, Leuchs G and Polzik E S (ed) 2007 *Quantum Information with Continuous Variables of Atoms and Light* (London: Imperial College Press)

[5] Shapiro J H 2008 Quantum optical communication *Lecture notes* http://ocw.mit.edu/courses/electrical-engineering-and-computer-science/6-453-quantum-optical-communication-fall-2008

[6] Kok P and Lovett B W 2010 *Introduction to Optical Quantum Information Processing* (Cambridge: Cambridge University Press)

Acknowledgements

I am very grateful to many people who helped me shape this project. Most of the material upon which this book is based originated from three events. First, from my close collaboration with Raúl García-Patrón, Nicolas J Cerf, and Jeffrey H Shapiro, whose guidance and clarifying explanations gave me the motivation to start writing the precursor of the original manuscript in 2010. I am especially indebted to Raúl and Nicolas for inviting me into the topics they had already been working on and for being the kindest and most patient teachers. Second, from the inspiring interaction with my PhD advisors Germán J de Valcárcel and Eugenio Roldán, in particular when writing my dissertation. And finally, from the weeks I spent in Doha with Hyunchul Nha and his group in 2012, during which many parts of the original manuscript were refined, benefiting greatly from discussions with Ho-Joon Kim and Hyunchul himself.

A special mention is well deserved by my good friend Erez Zohar, who was not only kind enough to write my biography for the monograph, but also read thoroughly the whole manuscript, helping me to improve it a great deal. Along these lines, I also appreciate the effort made by those who took some time to read through the manuscript and give me valuable feedback. These include Yue Chang, Johannes Kofler, Sebastian Pina-Otey, Ho-Joon Kim, Anaelle Hertz, Raúl García-Patrón, Eliška Greplová, Germán J de Valcárcel, Eugenio Roldán, Tao Shi, Alejandro González-Tudela, Nikolas Perakis, Marta P Estarellas, Samuel Fernández, Juan Bermejo-Vega, and Na'ama Hallakoun.

I feel blessed for having always been able to grow scientifically in inspiring environments, surrounded by nice colleagues and collaborators that I admire. I want to extend my appreciation to all of them, especially to the ones with whom I have discussed and applied many of the concepts that appear in this book: Mari Carmen Bañuls, Claude Fabre, Chiara Molinelli, Giuseppe Patera, J Ignacio Cirac, Inés de Vega, Diego Porras, Géza Giedke, Maarten Van den Nest, Peter D Drummond, Juan José García-Ripoll, Joaquín Ruiz-Rivas, Oriol Romero-Isart, Fernando Pastawski, Martin Schuetz, Jaehak Lee, Jiyong Park, Chang-Woo Lee, Se-Wan Ji, Myungshik Kim, Tao Shi, Yue Chang, Hannes Pichler, Peter Zoller, Julio T Barreiro, Alejandro González-Tudela, Eliška Greplová, Christopher A Fuchs, Christoph Marquardt, Florian Marquardt, Peter Degenfeld-Schonburg, Michael J Hartmann, Fernando Jiménez, Sebastian Pina-Otey, Mónica Benito, Carlos Sánchez Muñoz, András Molnár, Katja Kustura, Giulia Ferrini, Tom Douce, Cosimo C Rusconi, Alessandro Farace, and Julian Roos.

I have been blessed as well with a loving family and group of friends who have always supported me all the way in whatever adventure I have set sail. I am especially grateful to my parents for imprinting on me the will to make a living out of the things I am passionate about (be it music, sports, or science).

I gave the final shape to this book in the Lost Weekend cafe in Munich, whose staff provided me with the perfect warm and creative environment. Particular thanks are due to Markus, Chris, Eitel, Jaemin, Audrey, Tony, Julietta, Twana, and Lissie.

Finally, I would like to thank the editors and publishers of this synthesis series, especially Jeffrey Uhlmann and Michael B Morgan with whom I have been in direct contact, for giving me this extraordinary opportunity.

Carlos Navarrete-Benlloch
Munich, November 2015.

Author's biography

Carlos Navarrete-Benlloch by Erez Zohar

Credit: Lisa Miletic

Carlos Navarrete-Benlloch (born 1983) was born and raised in Valencia, Spain, where he attended Universitat de València for a six-year BSc+MSc degree in theoretical physics. During his studies, he became fascinated by several physical fields. One of these fields, which he later continued studying during his PhD, was quantum optics.

Carlos completed his PhD degree in 2011 under the joint supervision of Eugenio Roldán and Germán J de Valcárcel, in the quantum and nonlinear optics group of the Universitat de València, focusing on the quantum properties of multi-mode optical parametric oscillators. During his PhD studies, Carlos also maintained his interest in various other topics in quantum optics, which he had the opportunity to work on and become acquainted with during several long-term visits he made to other universities and research institutions. These included the Max-Planck Institute of Quantum Optics (MPQ) in Garching, Germany, where he worked under the supervision of J Ignacio Cirac; the Swinburne University of Technology in Melbourne, Australia, under the supervision of Peter D Drummond; and the Massachusetts Institute of Technology, USA, under the supervision of Jeffrey H Shapiro.

After receiving his PhD, Carlos joined J Ignacio Cirac's theory division at MPQ as a postdoctoral researcher, including two years as the recipient of a postdoctoral research fellowship granted by the Alexander von Humboldt foundation. At MPQ, Carlos has mainly focused on dissipative quantum optics with nonlinear optical cavities, superconducting circuits, and optomechanical systems, as well as on quantum simulation with ultracold atoms, collaborating with several scientists inside and outside MPQ. In the field of quantum information with continuous variables, Carlos has contributed to the theory of non-Gaussian states and operations, as well as the characterization of the classical capacity of Gaussian channels, in collaboration with Raúl García-Patrón, Nicolas J Cerf, Jeffrey H Shapiro, and Seth Lloyd.

Apart from quantum optics and information, Carlos has studied and been interested in several other physical fields, such as gravitation, cosmology, quantum field theory, and particle physics. He has also studied modern guitar and is a founding member and composer of the Valencia-based progressive jazz band Versus Five.

An Introduction to the Formalism of Quantum Information
with Continuous Variables

Carlos Navarrete-Benlloch

Chapter 1

Quantum-mechanical description
of physical systems

The purpose of this first chapter is the introduction of the fundamental laws (axioms) of quantum mechanics as are used throughout the book. The quantum framework is far from being intuitive, but somehow feels reasonable (and even inevitable) once one understands the context in which it was created, specifically: (i) the theories that were used to describe physical systems prior to its development, (ii) the experiments which did not fit in this context, and (iii) the mathematical language that accommodates the new quantum formulation of physical phenomena. We will thus briefly review this context prior to introducing and discussing the axioms forming the quantum-mechanical framework.

1.1 Classical mechanics

In this section we go through a brief introduction to classical mechanics[1], with emphasis on analyzing the Hamiltonian formalism and how it treats observable magnitudes. We will see that a proper understanding of this formalism will make the transition to quantum mechanics more natural.

1.1.1 The Lagrangian formalism

In classical mechanics the state of a system is specified by the position of its constituent particles at all times, $\mathbf{r}_j(t) = [x_j(t), y_j(t), z_j(t)]$ with $j = 1, 2, ..., N$, N being the number of particles. Defining the kinetic momentum of the particles as $\mathbf{P}_j = m_j \dot{\mathbf{r}}_j$ (m_j is the mass of particle j), the evolution of the system is found from a

---

[1] For a more in depth text I recommend Goldstein's book [1], as well as Greiner's books [2, 3], or that by Hand and Finch [4].

set of initial positions and velocities by solving the Newton equations of motion $\dot{\mathbf{P}}_j = \mathbf{F}_j$, $\mathbf{F}_j$ being the forces acting on particle j.

Most physical systems have further constraints that have to be fulfilled (for example, the distance between the particles of a rigid body cannot change with time, that is, $|\mathbf{r}_j - \mathbf{r}_l| = const$), and therefore the positions $\{\mathbf{r}_j\}_{j=1,...,N}$ are no longer independent, which makes the Newton equations difficult to solve. This calls for a new simpler theoretical framework: the so-called *analytical mechanics*. In the following we review this framework, but assuming, for simplicity, that the constraints are *holonomic*[2] and *scleronomous*[3], meaning that they can always be written in the form $f(\mathbf{r}_1, ..., \mathbf{r}_N) = 0$.

In analytical mechanics the state of the system at any time is specified by a vector $\mathbf{q}(t) = [q_1(t), q_2(t), ..., q_n(t)]$. n is the number of degrees of freedom of the system (the total number of coordinates, $3N$, minus the number of constraints), and the q_j are called the *generalized coordinates* of the system, which are compatible with the constraints and related to the usual coordinates of the particles through some smooth functions $\mathbf{q}(\{\mathbf{r}_j\}_{j=1,...,N}) \Leftrightarrow \{\mathbf{r}_j(\mathbf{q})\}_{j=1,...,N}$. The space formed by the generalized coordinates is called *coordinate space*, and $\mathbf{q}(t)$ describes a *trajectory* on it.

The basic object in analytical mechanics is the *Lagrangian*, $L[\mathbf{q}(t), \dot{\mathbf{q}}(t), t]$, which is a function of the generalized coordinates and velocities, and can even have some explicit time dependence. In general, the Lagrangian must be built based on general principles such as symmetries. However, if the forces acting on the particles of the system are conservative, that is, $\mathbf{F}_j = \nabla_j V[\{\mathbf{r}_l\}_{l=1,...,N}] = (\partial_{x_j} V, \partial_{y_j} V, \partial_{z_j} V)$ for some *potential* $V[\{\mathbf{r}_l\}_{l=1,...,N}]$, one can choose a Lagrangian with the simple form $L = T(\dot{\mathbf{q}}, \mathbf{q}) - V(\mathbf{q})$, $T(\dot{\mathbf{q}}, \mathbf{q}) = \sum_{j=1}^{N} m_j \dot{\mathbf{r}}_j^2(\mathbf{q})/2$ being the kinetic energy of the system and $V(\mathbf{q}) = V[\{\mathbf{r}_l(\mathbf{q})\}_{l=1,...,n}]$. The dynamical equations of the system are then formulated as a *variational principle* on the *action*

$$S = \int_{t_1}^{t_2} \mathrm{d}t L[\mathbf{q}(t), \dot{\mathbf{q}}(t), t], \tag{1.1}$$

by asking the trajectory of the system $\mathbf{q}(t)$ between two fixed points $\mathbf{q}(t_1)$ and $\mathbf{q}(t_2)$ to be such that the action is an extremal, $\delta S = 0$. From this principle, it is straightforward to arrive at the well known Euler–Lagrange equations

$$\frac{\partial L}{\partial q_j} - \frac{\mathrm{d}}{\mathrm{d}t}\frac{\partial L}{\partial \dot{q}_j} = 0, \tag{1.2}$$

which are a set of second-order differential equations for the generalized coordinates $\mathbf{q}$, and together with the conditions $\mathbf{q}(t_1)$ and $\mathbf{q}(t_2)$ provide the trajectory $\mathbf{q}(t)$.

---

[2] A constraint is *holonomic* when it can be written as $f(\mathbf{r}_1, ..., \mathbf{r}_N, t) = 0$. Non-holonomic constraints correspond, for example, to the boundary imposed by a wall that particles cannot cross, which is usually expressed in terms of inequalities [1], and requires a much more careful treatment.
[3] A constraint is called *scleronomous* when it does not depend explicitly on time. Time-dependent constraints are called *rheonomous* and correspond, for example, to a situation in which the motion of the particles is restricted to a moving surface or curve [1].

1.1.2 The Hamiltonian formalism

As we have seen, the Euler-Lagrange equations are a set of second-order differential equations which allows us to find the trajectory $\mathbf{q}(t)$ in coordinate space. We could reduce the order of the differential equations by taking the velocities $\dot{\mathbf{q}}$ as dynamical variables, arriving then at a set of $2n$ first-order differential equations. This is, however, a very naïve way of reducing the order, which leads to a non-symmetric system of equations for $\mathbf{q}$ and $\dot{\mathbf{q}}$. In this section we review Hamilton's approach to analytical mechanics, which leads to a symmetric-like first-order system of equations and will play a major role in understanding the transition from classical to quantum mechanics.

Instead of using the velocities, the Hamiltonian formalism considers the *generalized momenta*

$$p_j = \frac{\partial L}{\partial \dot{q}_j}, \tag{1.3}$$

as the dynamical variables. Note that this definition establishes a relation between these generalized momenta and the velocities $\dot{\mathbf{q}}(\mathbf{q}, \mathbf{p}) \Leftrightarrow \mathbf{p}(\mathbf{q}, \dot{\mathbf{q}})$. Note also that when the usual Cartesian coordinates of the system's particles are taken as the generalized coordinates, these momenta coincide with those of Newton's approach.

The theory is then built in terms of a new object called the *Hamiltonian*, which is defined as a Legendre transform of the Lagrangian,

$$H(\mathbf{q}, \mathbf{p}) = \mathbf{p}\dot{\mathbf{q}}(\mathbf{q}, \mathbf{p}) - L[\mathbf{q}, \dot{\mathbf{q}}(\mathbf{q}, \mathbf{p}), t], \tag{1.4}$$

and coincides with the total energy[4] for conservative systems with scleronomous constraints, that is, $H(\mathbf{q}, \mathbf{p}) = T(\mathbf{q}, \mathbf{p}) + V(\mathbf{q})$, with $T(\mathbf{q}, \mathbf{p}) = T[\mathbf{q}, \dot{\mathbf{q}}(\mathbf{q}, \mathbf{p})]$. Differentiating this expression and using the Euler–Lagrange equations (or using again the variational principle on the action), it is then straightforward to obtain the equations of motion for the generalized coordinates and momenta (the *canonical equations*),

$$\dot{q}_j = \frac{\partial H}{\partial p_j} \qquad \text{and} \qquad \dot{p}_j = -\frac{\partial H}{\partial q_j}, \tag{1.5}$$

which together with some initial conditions $\{\mathbf{q}(t_0), \mathbf{p}(t_0)\}$ allow us to find the trajectory $\{\mathbf{q}(t), \mathbf{p}(t)\}$ in the space formed by the generalized coordinates and momenta, which is known as *phase space*.

Another important object in the Hamiltonian formalism is the *Poisson bracket*. Given two functions of the coordinates and momenta $F(\mathbf{q}, \mathbf{p})$ and $G(\mathbf{q}, \mathbf{p})$, their Poisson bracket is defined as

$$\{F, G\} = \sum_{j=1}^{n} \frac{\partial F}{\partial q_j} \frac{\partial G}{\partial p_j} - \frac{\partial F}{\partial p_j} \frac{\partial G}{\partial q_j}. \tag{1.6}$$

The importance of this object is reflected in the fact that the evolution equation of any quantity $g(\mathbf{q}, \mathbf{p}, t)$ can be written as

---

[4] The general conditions under which the Hamiltonian coincides with the system's energy can be found in [1].

$$\frac{dg}{dt} = \{g, H\} + \frac{\partial g}{\partial t},\tag{1.7}$$

and hence, if the quantity does not depend explicitly on time and its Poisson bracket with the Hamiltonian is zero, it is a *constant of motion*.

Of particular importance for the transition to quantum mechanics are the *canonical Poisson brackets*, that is, the Poisson brackets of the coordinates and momenta,

$$\{q_j, p_l\} = \delta_{jl}, \qquad \{q_j, q_l\} = \{p_j, p_l\} = 0,\tag{1.8}$$

which define the mathematical structure of phase space.

1.1.3 Observables and their mathematical structure

In this last section concerning classical mechanics we will discuss the mathematical structure in which observables are embedded within the Hamiltonian formalism. We will see that the mathematical objects corresponding to physical observables form a well defined mathematical structure, a real Lie algebra. Moreover, the position and momentum will be shown to be the generators of a particular Lie group, the Heisenberg group. Understanding this internal structure of *classical observables* will give us the chance to introduce the quantum description of observables in a reasonable way. Let us start by defining the concept of Lie algebra.

A *real Lie algebra* is a real vector space[5] $\mathcal{L}$ equipped with an additional operation, the *Lie product*, which takes two vectors f and g from $\mathcal{L}$, to generate another vector also in $\mathcal{L}$ denoted by[6] $\{f, g\}$. This operation must satisfy the following properties:

1. $\{f, g + h\} = \{f, g\} + \{f, h\}$ (linearity)
2. $\{f, f\} = 0 \overset{\text{together}}{\underset{\text{with 1}}{\Longrightarrow}} \{f, g\} = -\{g, f\}$ (anticommutativity)
3. $\{f, \{g, h\}\} + \{g, \{h, f\}\} + \{h, \{f, g\}\} = 0$ (Jacobi identity).

Hence, in essence a real Lie algebra is a vector space equipped with a linear, non-commutative, non-associative product. They have been a subject of study for many years, and now we know a lot about the properties of these mathematical structures. They appear in many branches of physics and geometry, in particular connected to continuous symmetry transformations, whose associated mathematical structures are actually called *Lie groups*. In particular, it is possible to show that given any Lie group with p parameters (such as, e.g., the three-parameter groups of translations or rotations in real space), any transformation onto the system in which it is acting can be generated from a set of p elements of a Lie

---

[5] The concept of complex vector space is defined in the next section. The definition of a *real* vector space is the same, but replacing by real numbers the complex numbers that appear in the definitions there.

[6] Note that we are using the same notation for the general definition of the Lie product and for the Poisson bracket, which is a particular case of Lie product as we will learn shortly.

algebra $\{g_1, g_2, ..., g_p\}$, called the *generators* of the Lie group, which satisfy some particular relations

$$\{g_j, g_k\} = \sum_{l=1}^{p} c_{jkl}g_l. \tag{1.9}$$

These relations are called the *algebra-group relations*, and the *structure constants* c_{jkl} are characteristic of the particular Lie group (for example, the generators of translations and rotations in real space are the momenta and angular momenta, respectively, and the corresponding structure constants are $c_{jkl} = 0$ for the translation group and $c_{jkl} = \epsilon_{jkl}$ for the rotation group[7]).

Coming back to the Hamiltonian formalism, we start by noting that *observables*, being *measurable* quantities, must be given by continuous, real functions in phase space. Hence they form a real vector space with respect to the usual addition of functions and multiplication of a function by a real number. Also appearing naturally in the formalism is a linear, non-commutative, non-associative operation between phase-space functions, the Poisson bracket, which applied to real functions gives another real function. It is easy to see that the Poisson bracket satisfies all the requirements of a Lie product, and hence, observables form a Lie algebra within the Hamiltonian formalism.

Moreover, the canonical Poisson brackets (1.8) show that the generalized coordinates $\mathbf{q}$ and momenta $\mathbf{p}$, together with the identity in phase space, satisfy particular algebra-group relations, namely[8] $\{q_j, p_k\} = \delta_{jk}1$ and $\{q_j, 1\} = \{p_j, 1\} = \{1, 1\} = 0$, and hence can be seen as the generators of a Lie group. This group is known as the *Heisenberg group*, and was introduced by Weyl when trying to prove the equivalence between the Schrödinger and Heisenberg pictures of quantum mechanics (which we will learn about later). It was later shown to have connections with the symplectic group, which is the basis of many physical theories. Note that we could have taken the Poisson brackets between the angular momenta associated with the possible rotations in the system of particles (which are certainly far more intuitive transformations than the one related to the Heisenberg group) as the fundamental ones. However, we have chosen the Lie algebra associated with the Heisenberg group just because it deals directly with position and momentum, allowing for a simpler connection to quantum mechanics.

---

[7] $\epsilon_{j_1 j_2 \, ... \, j_M}$ with all the subindices going from 1 to M is the Levi-Civita symbol in dimension M, which has $\epsilon_{12 \, ... \, M} = 1$ and is completely antisymmetric, that is, changes its sign after permutation of any pair of indices.
[8] Ordering the generators as $\{\mathbf{q}, \mathbf{p}, 1\}$, the structure constants associated with these algebra-group relations are explicitly

$$c_{jkl} = \begin{cases} \Omega_{jk}\delta_{l,2n+1} & j, k = 1, 2, ..., 2n \\ 0 & j = 2n + 1 \text{ or } k = 2n + 1, \end{cases} \tag{1.10}$$

where $\Omega = \begin{pmatrix} 0_{n\times n} & I_{n\times n} \\ -I_{n\times n} & 0_{n\times n} \end{pmatrix}$, with $I_{n\times n}$ and $0_{n\times n}$ the $n \times n$ identity and null matrices, respectively.

Therefore, we arrive at the main conclusion of this review of classical mechanics: the mathematical framework of Hamiltonian mechanics associates physical observables with elements of a Lie algebra, with the phase-space coordinates themselves being the generators of the Heisenberg group.

Maintaining this structure for observables will help us introduce the laws of quantum mechanics in a coherent way.

1.2 The mathematical language of quantum mechanics

Just as classical mechanics is formulated in terms of the mathematical language of differential calculus and its extensions, quantum mechanics takes linear algebra (and Hilbert spaces in particular) as its fundamental grammar. In this section we will review the concept of Hilbert space, and discuss the properties of some operators which play important roles in the formalism of quantum information.

1.2.1 Finite-dimensional Hilbert spaces

In essence, a Hilbert space is a *complex vector space* in which an *inner product* is defined. Let us define first these terms as we use them in this book.

A *complex vector space* is a set $\mathcal{V}$, whose elements will be called *vectors* or *kets* and will be denoted by $\{|a\rangle, |b\rangle, |c\rangle, ...\}$ (a, b, and c may correspond to any suitable label), in which the following two operations are defined: the *vector addition*, which takes two vectors $|a\rangle$ and $|b\rangle$ and creates a new vector inside $\mathcal{V}$ denoted by $|a\rangle + |b\rangle$; and the *multiplication by a scalar*, which takes a complex number $\alpha \in \mathbb{C}$ (in this section Greek letters will represent complex numbers) and a vector $|a\rangle$ to generate a new vector in $\mathcal{V}$ denoted by $\alpha|a\rangle$.

The following additional properties must be satisfied:
1. The vector addition is commutative and associative, that is, $|a\rangle + |b\rangle = |b\rangle + |a\rangle$ and $(|a\rangle + |b\rangle) + |c\rangle = |a\rangle + (|b\rangle + |c\rangle)$.
2. There exists a null vector $|null\rangle$ such that $|a\rangle + |null\rangle = |a\rangle$
3. $\alpha(|a\rangle + |b\rangle) = \alpha|a\rangle + \alpha|b\rangle$
4. $(\alpha + \beta)|a\rangle = \alpha|a\rangle + \beta|a\rangle$
5. $(\alpha\beta)|a\rangle = \alpha(\beta|a\rangle)$
6. $1|a\rangle = |a\rangle$.

From these properties it can be proved that the null vector is unique, and can be built from any vector $|a\rangle$ as $0|a\rangle$; hence, in the following we denote it simply by $|null\rangle \equiv 0$. It is also readily proved that any vector $|a\rangle$ has a unique *antivector* $|-a\rangle$ such that $|a\rangle + |-a\rangle = 0$, which is given by $(-1)|a\rangle$ or simply $-|a\rangle$.

An *inner product* is an additional operation defined in the complex vector space $\mathcal{V}$, which takes two vectors $|a\rangle$ and $|b\rangle$ and associates them with a complex number. It will be denoted by $\langle a|b\rangle$ or sometimes also by $(|a\rangle, |b\rangle)$, and must satisfy the following properties:
1. $\langle a|a\rangle > 0$ if $|a\rangle \neq 0$
2. $\langle a|b\rangle = \langle b|a\rangle^*$
3. $(|a\rangle, \alpha|b\rangle) = \alpha\langle a|b\rangle$
4. $(|a\rangle, |b\rangle + |c\rangle) = \langle a|b\rangle + \langle a|c\rangle$.

The following additional properties can be proved from these ones:
- $\langle null|null \rangle = 0$
- $(\alpha|a\rangle, |b\rangle) = \alpha^* \langle a|b \rangle$
- $(|a\rangle + |b\rangle, |c\rangle) = \langle a|c \rangle + \langle b|c \rangle$
- $|\langle a|b \rangle|^2 \leqslant \langle a|a \rangle \langle b|b \rangle$ (Cauchy-Schwarz).

Note that for any vector $|a\rangle$, one can define the object $\langle a| \equiv (|a\rangle, \cdot)$, which will be called a *dual vector* or a *bra*, and which takes a vector $|b\rangle$ to generate the complex number $(|a\rangle, |b\rangle) \in \mathbb{C}$. It can be proved that the set formed by all the dual vectors corresponding to the elements in $\mathcal{V}$ is also a vector space, which will be called the *dual space* and will be denoted by $\mathcal{V}^+$. Within this picture, the inner product can be seen as an operation which takes a bra $\langle a|$ and a ket $|b\rangle$ to generate the complex number $\langle a|b \rangle$, a *bracket*. This whole *bra-c-ket* notation is due to Dirac [5].

In the following we assume that any time a bra $\langle a|$ is applied to a ket $|b\rangle$, the complex number $\langle a|b \rangle$ is formed, so that objects such as $|b\rangle\langle a|$ generate kets when applied to kets from the left, $(|b\rangle\langle a|)|c\rangle = (\langle a|c\rangle)|b\rangle$, and bras when applied to bras from the right, $\langle c|(|b\rangle\langle a|) = (\langle c|b\rangle)\langle a|$. Technically, $|b\rangle\langle a|$ is called an *outer product*.

A vector space equipped with an inner product is called a *Euclidean space* [6]. In the following we give some important definitions and properties which are needed in order to understand the concept of Hilbert space:

- The vectors $\{|a_1\rangle, |a_2\rangle, ..., |a_m\rangle\}$ are said to be *linearly independent* if the relation $\alpha_1|a_1\rangle + \alpha_2|a_2\rangle + \cdots + \alpha_m|a_m\rangle = 0$ is satisfied only for $\alpha_1 = \alpha_2 = \cdots = \alpha_m = 0$, as otherwise one of them can be written as a linear combination of the rest.
- The *dimension* of the vector space is defined as the maximum number of linearly independent vectors that can be found in the space, and can be finite or infinite.
- If the dimension of a Euclidean space is $d < \infty$, it is always possible to build a set of d orthonormal vectors $E = \{|e_j\rangle\}_{j=1,2,...,d}$ satisfying $\langle e_j|e_l \rangle = \delta_{jl}$, such that any other vector $|a\rangle$ can be written as a linear superposition of them, that is, $|a\rangle = \sum_{j=1}^{d} a_j|e_j\rangle$, the a_j being some complex numbers. This set is called an *orthonormal basis* of the Euclidean space $\mathcal{V}$, and the coefficients a_j of the expansion can be found as $a_j = \langle e_j|a \rangle$. The column formed with the expansion coefficients, which is denoted by $\mathrm{col}(a_1, a_2, ..., a_d)$, is called a *representation* of the vector $|a\rangle$ in the basis E. Note that the set $E^+ = \{\langle e_j|\}_{j=1,2,...,d}$ is an orthonormal basis in the dual space $\mathcal{V}^+$, so that any bra $\langle a|$ can be expanded then as $\langle a| = \sum_{j=1}^{d} a_j^* \langle e_j|$. The representation of the bra $\langle a|$ in the basis E corresponds to the row formed by its expansion coefficients, and is denoted by $(a_1^*, a_2^*, ..., a_n^*)$. Note that if the representation of $|a\rangle$ is seen as a $d \times 1$ matrix, the representation of $\langle a|$ can be obtained as its $1 \times d$ conjugate-transpose matrix.

  Note finally that the inner product of two vectors $|a\rangle$ and $|b\rangle$ reads $\langle a|b \rangle = \sum_{j=1}^{d} a_j^* b_j$ when represented in the same basis, which is the matrix product of the representations of $\langle a|$ and $|b\rangle$.

For finite dimensions, a Euclidean space is a *Hilbert space*. However, in most applications of quantum mechanics (and certainly in continuous-variable quantum information), one has to deal with infinite-dimensional vector spaces. We will treat them after the following section.

1.2.2 Linear operators in finite dimensions

We now discuss the concept of linear operator, as well as analyze the properties of some important classes of operators. Only finite-dimensional Hilbert spaces are considered in this section, and we will generalize the discussion to infinite-dimensional Hilbert spaces in the next section.

We are interested in maps $\hat{L}$ (operators are denoted by a 'hat' throughout) which associate to any vector $|a\rangle$ of a Hilbert space $\mathcal{H}$ another vector denoted by $\hat{L}|a\rangle$ in the same Hilbert space. If the map satisfies

$$\hat{L}(\alpha|a\rangle + \beta|b\rangle) = \alpha\hat{L}|a\rangle + \beta\hat{L}|b\rangle, \tag{1.11}$$

then it is called a *linear operator*. For our purposes this is the only class of interesting operators, and hence we will simply call them *operators* in the following. Before discussing the properties of some important classes of operators, we need some definitions:

- Given an orthonormal basis $E = \{|e_j\rangle\}_{j=1,2,\ldots,d}$ in a Hilbert space $\mathcal{H}$ with dimension $d < \infty$, any operator $\hat{L}$ has a matrix representation. While bras and kets are represented by $d \times 1$ and $1 \times d$ matrices (rows and columns), respectively, an operator $\hat{L}$ is represented by a $d \times d$ matrix with *elements* $L_{jl} = (|e_j\rangle, \hat{L}|e_l\rangle) \equiv \langle e_j|\hat{L}|e_l\rangle$. An operator $\hat{L}$ can then be expanded in terms of the basis E as $\hat{L} = \sum_{j,l=1}^{d} L_{jl}|e_j\rangle\langle e_l|$. It follows that the representation of the vector $|b\rangle = \hat{L}|a\rangle$ is just the matrix multiplication of the representation of $\hat{L}$ by the representation of $|a\rangle$, that is, $b_j = \sum_{l=1}^{d} L_{jl}a_l$.

- The *addition* and *multiplication* of two operators $\hat{L}$ and $\hat{K}$, denoted by $\hat{L} + \hat{K}$ and $\hat{L}\hat{K}$, respectively, are defined by their action onto any vector $|a\rangle$: $(\hat{L} + \hat{K})|a\rangle = \hat{L}|a\rangle + \hat{K}|a\rangle$ and $\hat{L}\hat{K}|a\rangle = \hat{L}(\hat{K}|a\rangle)$. It follows that the representation of the addition and the product are, respectively, the sum and the multiplication of the corresponding matrices, that is, $(\hat{L} + \hat{K})_{jl} = L_{jl} + K_{jl}$ and $(\hat{L}\hat{K})_{jl} = \sum_{k=1}^{d} L_{jk}K_{kl}$.

- Note that while the addition is commutative, the product in general is not. This leads us to the notion of the *commutator*, defined for two operators $\hat{L}$ and $\hat{K}$ as $[\hat{L}, \hat{K}] = \hat{L}\hat{K} - \hat{K}\hat{L}$. When $[\hat{L}, \hat{K}] = 0$, we say that the operators *commute*.

- Given an operator $\hat{L}$, its *trace* is defined as the sum of the diagonal elements of its matrix representation, that is, $\mathrm{tr}\{\hat{L}\} = \sum_{j=1}^{d} L_{jj}$. It may seem that this definition is basis-dependent, as in general the elements L_{jj} are different in different bases. However, we will see later that the trace is invariant under any change of basis.

The trace has two important properties. It is *linear* and *cyclic*, that is, given two operators $\hat{L}$ and $\hat{K}$, $\mathrm{tr}\{\hat{L} + \hat{K}\} = \mathrm{tr}\{\hat{L}\} + \mathrm{tr}\{\hat{K}\}$ and $\mathrm{tr}\{\hat{L}\hat{K}\} = \mathrm{tr}\{\hat{K}\hat{L}\}$, as is trivial to prove.

- Given an operator $\hat{L}$, we define its *determinant* as the determinant of its matrix representation, that is, $\det\{\hat{L}\} = \sum_{j_1,j_2,\ldots,j_d=1}^{d} \epsilon_{j_1 j_2 \ldots j_d} L_{1j_1} L_{2j_2} \ldots L_{dj_d}$. Just as the trace, we will see that it does not depend on the basis used to represent the operator.

  The determinant is a multiplicative map, that is, given two operators $\hat{L}$ and $\hat{K}$, the determinant of the product is the product of the determinants, $\det\{\hat{L}\hat{K}\} = \det\{\hat{L}\}\det\{\hat{K}\}$.

- We say that a vector $|l\rangle$ is an *eigenvector* of an operator $\hat{L}$ if there exists a $\lambda \in \mathbb{C}$ (called its associated *eigenvalue*) such that $\hat{L}|l\rangle = \lambda|l\rangle$. The set of all the eigenvalues of an operator is called its *spectrum*.

We can now move on to describe some classes of operators which play important roles in quantum mechanics.

The identity operator. The *identity operator*, denoted by $\hat{I}$, is defined as the operator which maps any vector onto itself. Its representation in any basis is then $I_{jl} = \delta_{jl}$, so that it can be expanded as

$$\hat{I} = \sum_{j=1}^{d} |e_j\rangle\langle e_j|. \qquad (1.12)$$

This expression is known as the *completeness relation* of the basis E; alternatively, it is said that the set E forms a *resolution of the identity*.

Note that the expansion of a vector $|a\rangle$ and its dual $\langle a|$ in the basis E is obtained just by application of the completeness relation from the left and the right, respectively. Similarly, the expansion of an operator $\hat{L}$ is obtained by application of the completeness relation both from the right and the left at the same time.

The inverse of an operator. The *inverse* of an operator $\hat{L}$, denoted by $\hat{L}^{-1}$, is defined as that satisfying $\hat{L}^{-1}\hat{L} = \hat{L}\hat{L}^{-1} = \hat{I}$. Not every operator has an inverse. An inverse exists if and only if the operator does not have a zero eigenvalue, or, equivalently, when $\det\{\hat{L}\} \neq 0$.

An operator function. Consider a real, analytic function $f(x)$ which can be expanded in powers of x as $f(x) = \sum_{m=0}^{\infty} f_m x^m$. Given an operator $\hat{L}$, we define the *operator function* $\hat{f}(\hat{L}) = \sum_{m=0}^{\infty} f_m \hat{L}^m$, where $\hat{L}^m$ means the product of $\hat{L}$ with itself m times.

The adjoint of an operator. Given an operator $\hat{L}$, we define its *adjoint*, and denote it by $\hat{L}^{\dagger}$, as that satisfying $(|a\rangle, \hat{L}|b\rangle) = (\hat{L}^{\dagger}|a\rangle, |b\rangle)$ for any two vectors $|a\rangle$ and $|b\rangle$. Note that the representation of $\hat{L}^{\dagger}$ corresponds to the conjugate transpose of the matrix representing $\hat{L}$, that is, $(\hat{L}^{\dagger})_{jl} = L_{lj}^{*}$. Note also that the adjoint of a product of two operators $\hat{K}$ and $\hat{L}$ is given by $(\hat{K}\hat{L})^{\dagger} = \hat{L}^{\dagger}\hat{K}^{\dagger}$.

Self-adjoint operators. We say that $\hat{H}$ is *self-adjoint* when it coincides with its adjoint, that is, $\hat{H} = \hat{H}^{\dagger}$. A property of major importance for the construction of the laws of quantum mechanics is that the spectrum $\{h_j\}_{j=1,2,...,d}$ of a self-adjoint operator is real. Moreover, its associated eigenvectors[9] $\{|h_j\rangle\}_{j=1,2,...,d}$ form an orthonormal basis of the Hilbert space.

The representation of any operator function $\hat{f}(\hat{H})$ in the *eigenbasis* of $\hat{H}$ is then $[\hat{f}(\hat{H})]_{jl} = f(h_j)\delta_{jl}$, from which follows

$$\hat{f}(\hat{H}) = \sum_{j=1}^{d} f(h_j)|h_j\rangle\langle h_j|. \tag{1.13}$$

This result is known as the *spectral theorem*.

Unitary operators. We say that $\hat{U}$ is a *unitary operator* when $\hat{U}^\dagger = \hat{U}^{-1}$. The interest of this class of operators is that they preserve inner products, that is, for any two vectors $|a\rangle$ and $|b\rangle$ the inner product $(\hat{U}|a\rangle, \hat{U}|b\rangle)$ coincides with $\langle a|b\rangle$. Moreover, it is possible to show that given two orthonormal bases $E = \{|e_j\rangle\}_{j=1,2,...,d}$ and $E' = \{|e_j'\rangle\}_{j=1,2,...,d}$, there exists a unique unitary matrix $\hat{U}$ which connects them as $\{|e_j'\rangle = \hat{U}|e_j\rangle\}_{j=1,2,...,d}$, and then any basis of the Hilbert space is unique up to a unitary transformation.

We can now prove that both the trace and the determinant of an operator are basis-independent. Let us denote by $\mathrm{tr}\{\hat{L}\}_E$ the trace of an operator $\hat{L}$ in the basis E. The trace of this operator in the transformed basis can be written then as $\mathrm{tr}\{\hat{L}\}_{E'} = \mathrm{tr}\{\hat{U}^\dagger \hat{L}\hat{U}\}_E$, which, using the cyclic property of the trace and the unitarity of $\hat{U}$, is rewritten as $\mathrm{tr}\{\hat{U}\hat{U}^\dagger \hat{L}\}_E = \mathrm{tr}\{\hat{L}\}_E$, proving that the trace is equal in both bases. Similarly, in the case of the determinant we have $\det\{\hat{L}\}_{E'} = \det\{\hat{U}^\dagger \hat{L}\hat{U}\}_E$, which using the multiplicative property of the determinant is rewritten as $\det\{\hat{U}^\dagger\}_E \det\{\hat{L}\}_E \det\{\hat{U}\}_E = \det\{\hat{L}\}_E$, where we have used $\det\{\hat{U}^\dagger\}_E \det\{\hat{U}\}_E = 1$ as follows from $\hat{U}^\dagger\hat{U} = \hat{I}$.

Note finally that a unitary operator $\hat{U}$ can always be written as the complex exponential of a self-adjoint operator $\hat{H}$, that is, $\hat{U} = \exp(i\hat{H})$.

Projection operators. In general, any self-adjoint operator $\hat{P}$ satisfying $\hat{P}^2 = \hat{P}$ is called a *projector*. We are interested only in those projectors which can be written as the outer product of a vector $|a\rangle$ with itself, that is, rank-1 projectors[10] $\hat{P}_a = |a\rangle\langle a|$. When applied to a vector $|b\rangle$, this is *projected* along the 'direction' of $|a\rangle$ as $\hat{P}_a|b\rangle = \langle a|b\rangle|a\rangle$.

Note that given an orthonormal basis E, we can use the projectors $\hat{P}_j = |e_j\rangle\langle e_j|$ to extract the components of a vector $|a\rangle$ as $\hat{P}_j|a\rangle = a_j|e_j\rangle$. Note also that the completeness and orthonormality of the basis E implies that $\sum_{j=1}^{d}\hat{P}_j = \hat{I}$ and $\hat{P}_j\hat{P}_l = \delta_{jl}\hat{P}_j$, respectively.

---

[9] For simplicity, we will assume that the spectrum of any operator is non-degenerate, that is, all the eigenvectors possess a distinctive eigenvalue.
[10] The term 'rank' refers to the number of non-zero eigenvalues.

Density operators. A self-adjoint operator $\hat{\rho}$ is called a *density operator* when it has unit trace and it is *positive semidefinite*, that is, $\langle a|\hat{\rho}|a\rangle \geqslant 0$ for any vector $|a\rangle$.

The interesting property of density operators is that they 'contain' probability distributions in the diagonal of its representation. To see this just note that given an orthonormal basis E, the self-adjointness and positivity of $\hat{\rho}$ ensure that all its diagonal elements $\{\rho_{jj}\}_{j=1,2,...,d}$ are either positive or zero, that is, $\rho_{jj} \geqslant 0 \; \forall j$, while the unit trace makes them satisfy $\sum_{j=1}^{d} \rho_{jj} = 1$. Hence, the diagonal elements of a density operator have all the properties required by a *probability distribution*.

1.2.3 Generalization to infinite dimensions

Unfortunately, not all the previous concepts and objects that we have introduced for the finite-dimensional case are trivially generalized to infinite dimensions. In this section we discuss this generalization.

The first problem that we meet when dealing with infinite-dimensional Euclidean spaces is that the existence of a basis $\{|e_j\rangle\}_{j=1,2,...}$ in which any other vector can be represented as $|a\rangle = \sum_{j=1}^{\infty} a_j |e_j\rangle$ is not granted. The class of infinite-dimensional Euclidean spaces in which these infinite but countable bases exist are called *separable Hilbert spaces*, and are the ones relevant for the quantum description of physical systems.

The conditions which ensure that an infinite-dimensional Euclidean space is indeed a Hilbert space[11] can be found in, for example, reference [6]. Here we just want to stress that, quite intuitively, any infinite-dimensional Hilbert space[12] is *isomorphic* to the space called $l^2(\infty)$, which is formed by the column vectors $|a\rangle = \mathrm{col}(a_1, a_2,...)$ where the set $\{a_j \in \mathbb{C}\}_{j=1,2,...}$ satisfies the restriction $\sum_{j=1}^{\infty} |a_j|^2 < \infty$, and has the operations $|a\rangle + |b\rangle = \mathrm{col}(a_1 + b_1, a_2 + b_2,...)$, $\alpha|a\rangle = \mathrm{col}(\alpha a_1, \alpha a_2,...)$, and $\langle a|b\rangle = \sum_{j=1}^{\infty} a_j^* b_j$.

Most of the previous definitions are directly generalized to Hilbert spaces by taking $d \to \infty$ (dual space, representations, operators,...). However, there is one crucial property of self-adjoint operators which does not hold in this case: their eigenvectors may not form an orthonormal basis of the Hilbert space. The remainder of this section is devoted to dealing with this problem.

---

[11] From now on we will assume that all the Hilbert spaces we refer to are 'separable', even if we do not write it explicitly.

[12] An example of infinite-dimensional complex Hilbert space consists in the vector space formed by the complex functions of real variable, say $|f\rangle = f(x)$ with $x \in \mathbb{R}$, with integrable square, that is

$$\int_{\mathbb{R}} \mathrm{d}x \, |f(x)|^2 < \infty, \tag{1.14}$$

together with the inner product

$$\langle g|f\rangle = \int_{\mathbb{R}} \mathrm{d}x g^*(x) f(x). \tag{1.15}$$

This Hilbert space is usually denoted by $L^2(x)$.

Just as in finite dimensions, given an infinite-dimensional Hilbert space $\mathcal{H}$, we say that one of its vectors $|d\rangle$ is an eigenvector of the self-adjoint operator $\hat{H}$ if $\hat{H}|d\rangle = \delta|d\rangle$, where $\delta \in \mathbb{R}$ is called its associated eigenvalue. Nevertheless, it can happen in infinite-dimensional spaces that some vector $|c\rangle$ not contained in $\mathcal{H}$ also satisfies the condition $\hat{H}|c\rangle = \chi|c\rangle$, in which case we call it a *generalized eigenvector*, χ being its *generalized eigenvalue*[13]. The set of all eigenvalues of the self-adjoint operator is called its *discrete* (or *point*) *spectrum* and it is a countable set, while the set of all its generalized eigenvalues is called its *continuous spectrum* and it is uncountable, that is, it forms a continuous set [6] (see also [7]).

In this monograph we only deal with two extreme cases: either the observable, say $\hat{H}$, has a pure discrete spectrum $\{h_j\}_{j=1,2,\ldots}$; or the observable, say $\hat{X}$, has a pure continuous spectrum $\{x\}_{x \in \mathbb{R}}$. It can be shown that in the first case the eigenvectors of the observable form an orthonormal basis of the Hilbert space, so that we can build a resolution of the identity as $\hat{I} = \sum_{j=1}^{\infty} |h_j\rangle\langle h_j|$, and proceed along the lines of the previous sections.

In the second case, the set of generalized eigenvectors cannot form a basis of the Hilbert space in the strict sense, as they do not form a countable set and do not even belong to the Hilbert space. Fortunately, there are still ways to treat the generalized eigenvectors of $\hat{X}$ 'as if' they were a basis of the Hilbert space. This idea was introduced by Dirac [5], who realized that normalizing the generalized eigenvectors as[14] $\langle x|y\rangle = \delta(x - y)$, one can define the following integral operator

$$\int_{\mathbb{R}} \mathrm{d}x |x\rangle\langle x| = \hat{I}_{\mathrm{c}}, \tag{1.19}$$

which acts as the identity onto the generalized eigenvectors, that is, $\hat{I}_{\mathrm{c}}|y\rangle = |y\rangle$. It is then assumed that $\hat{I}_{\mathrm{c}}$ coincides with the identity in $\mathcal{H}$, so that any other vector $|a\rangle$ or operator $\hat{L}$ defined in the Hilbert space can be expanded as

$$|a\rangle = \int_{\mathbb{R}} \mathrm{d}x\, a(x)|x\rangle \qquad \text{and} \qquad \hat{L} = \int_{\mathbb{R}^2} \mathrm{d}x\mathrm{d}y\, L(x, y)|x\rangle\langle y|, \tag{1.20}$$

where the elements $a(x) = \langle x|a\rangle$ and $L(x, y) = \langle x|\hat{L}|y\rangle$ of these *continuous representations* form complex functions defined in $\mathbb{R}$ and $\mathbb{R}^2$, respectively. From now on, we will call *continuous basis* to the set $\{|x\rangle\}_{x \in \mathbb{R}}$.

---

[13] In $\mathrm{L}^2(x)$ we have two simple examples of self-adjoint operators with eigenvectors not contained in $\mathrm{L}^2(x)$: the so-called $\hat{X}$ (*position*) and $\hat{P}$ (*momentum*), which, given an arbitrary vector $|f\rangle = f(x)$, act as $\hat{X}|f\rangle = xf(x)$ and $\hat{P}|f\rangle = -\mathrm{i}\partial_x f$, respectively. This is simple to see, as the equations

$$xf_X(x) = Xf_X(x) \qquad \text{and} \qquad -\mathrm{i}\partial_x f_P(x) = Pf_P(x), \tag{1.16}$$

have

$$f_X(x) = \delta(x - X) \qquad \text{and} \qquad f_P(x) = \exp(\mathrm{i}Px), \tag{1.17}$$

as solutions, which are not square-integrable, and hence do not belong to $\mathrm{L}^2(x)$.

[14] $\delta(x)$ is the so-called *Dirac-delta distribution* which is defined by the conditions

$$\int_{x_1}^{x_2} \mathrm{d}x\delta(x - y) = \begin{cases} 1 & \text{if } y \in [x_1, x_2] \\ 0 & \text{if } y \notin [x_1, x_2] \end{cases}. \tag{1.18}$$

Dirac introduced this continuous representation as a 'limit to the continuum' of the countable case. Even though this approach was very intuitive, it lacked mathematical rigor. Some decades after Dirac's proposal, Gel'fand showed how to generalize the concept of Hilbert space to include these generalized representations in full mathematical rigor [8]. The generalized spaces are called *rigged Hilbert spaces* (in which the algebra of Hilbert spaces joins forces with the theory of continuous probability distributions), and working on them it is possible to show that given any self-adjoint operator, one can use its eigenvectors and generalized eigenvectors to expand any vector of the Hilbert space, just as we did above. In other words, within the framework of rigged Hilbert spaces, one can prove the identity $\hat{I}_c = \hat{I}$ rigorously.

Note finally that given two vectors $|a\rangle$ and $|b\rangle$ of the Hilbert space, and a continuous basis $\{|x\rangle\}_{x\in\mathbb{R}}$, we can use their generalized representations to write their inner product as

$$\langle a|b\rangle = \int_{\mathbb{R}} \mathrm{d}x a^*(x)b(x). \qquad (1.21)$$

It is also easily proved that the trace of any operator $\hat{L}$ can be evaluated from its continuous representation on $\{|x\rangle\}_{x\in\mathbb{R}}$ as

$$\mathrm{tr}\{\hat{L}\} = \int_{\mathbb{R}} \mathrm{d}x L(x, x). \qquad (1.22)$$

This has important consequences for the properties of density operators, say $\hat{\rho}$ for the discussion which follows. We explained at the end of the last section that when represented on an orthonormal basis of the Hilbert space, its diagonal elements (which are real owing to its self-adjointness) can be seen as a probability distribution, because they satisfy $\sum_{j=1}^{\infty} \rho_{jj} = 1$ and $\rho_{jj} \geq 0$ $\forall j$. Similarly, because of its unit trace and positivity, the diagonal elements of its continuous representation satisfy $\int_{\mathbb{R}} \mathrm{d}x \rho(x, x) = 1$ and $\rho(x, x) \geq 0$ $\forall x$, and hence, the real function $\rho(x, x)$ can be seen as a *probability density function*.

1.2.4 Composite Hilbert spaces

In many moments of this monograph, we will find the need to associate a Hilbert space with a composite system, the Hilbert spaces of whose parts we know. In this section we show how to build a Hilbert space $\mathcal{H}$ starting from a set of Hilbert spaces $\{\mathcal{H}_A, \mathcal{H}_B, \mathcal{H}_C, \ldots\}$.

Let us start with only two Hilbert spaces $\mathcal{H}_A$ and $\mathcal{H}_B$ with dimensions d_A and d_B, respectively (which might be infinite); the generalization to an arbitrary number of Hilbert spaces is straightforward. Consider a vector space $\mathcal{V}$ with dimension $\dim(\mathcal{V}) = d_A \times d_B$. We define a map called the *tensor product* which associates to any pair of vectors $|a\rangle \in \mathcal{H}_A$ and $|b\rangle \in \mathcal{H}_B$ a vector in $\mathcal{V}$ which we denote by $|a\rangle \otimes |b\rangle \in \mathcal{V}$. This tensor product must satisfy the following properties:

1. $(|a\rangle + |b\rangle) \otimes |c\rangle = |a\rangle \otimes |c\rangle + |b\rangle \otimes |c\rangle$
2. $|a\rangle \otimes (|b\rangle + |c\rangle) = |a\rangle \otimes |b\rangle + |a\rangle \otimes |c\rangle$
3. $(\alpha|a\rangle) \otimes |b\rangle = |a\rangle \otimes (\alpha|b\rangle)$.

If we endorse the vector space $\mathcal{V}$ with the inner product $(|a\rangle \otimes |b\rangle, |c\rangle \otimes |d\rangle) = \langle a|c\rangle \langle b|d\rangle$, it is easy to show it becomes a Hilbert space, which in the following will be denoted by $\mathcal{H} = \mathcal{H}_A \otimes \mathcal{H}_B$. Given the bases $E_A = \{|e_j^A\rangle\}_{j=1,2,...,d_A}$ and $E_B = \{|e_j^B\rangle\}_{j=1,2,...,d_B}$ of the Hilbert spaces $\mathcal{H}_A$ and $\mathcal{H}_B$, respectively, a basis of the *tensor product Hilbert space* $\mathcal{H}_A \otimes \mathcal{H}_B$ can be built as $E = E_A \otimes E_B = \{|e_j^A\rangle \otimes |e_l^B\rangle\}_{l=1,2,...,d_B}^{j=1,2,...,d_A}$ (note that the notation after the first equality is symbolic).

We may use a more economic notation for the tensor product, namely $|a\rangle \otimes |b\rangle = |a, b\rangle$, except when the explicit tensor product symbol is needed for some special reason. With this notation the basis of the tensor product Hilbert space is written as $E = \{|e_j^A, e_l^B\rangle\}_{l=1,2,...,d_B}^{j=1,2,...,d_A}$.

The tensor product also maps operators acting on $\mathcal{H}_A$ and $\mathcal{H}_B$ to operators acting on $\mathcal{H}$. Given two operators $\hat{L}_A$ and $\hat{L}_B$ acting on $\mathcal{H}_A$ and $\mathcal{H}_B$, the *tensor product operator* $\hat{L} = \hat{L}_A \otimes \hat{L}_B$ is defined in $\mathcal{H}$ as that satisfying $\hat{L}|a, b\rangle = (\hat{L}_A|a\rangle) \otimes (\hat{L}_B|b\rangle)$ for any pair of vectors $|a\rangle \in \mathcal{H}_A$ and $|b\rangle \in \mathcal{H}_B$. When explicit subindices making reference to the Hilbert space on which operators act on are used, so that there is no room for confusion, we will use the shorter notations $\hat{L}_A \otimes \hat{L}_B = \hat{L}_A \hat{L}_B$, $\hat{L}_A \otimes \hat{I} = \hat{L}_A$, and $\hat{I} \otimes \hat{L}_B = \hat{L}_B$.

Note that the tensor product preserves the properties of the operators; for example, given two self-adjoint operators $\hat{H}_A$ and $\hat{H}_B$, unitary operators $\hat{U}_A$ and $\hat{U}_B$, or density operators $\hat{\rho}_A$ and $\hat{\rho}_B$, the operators $\hat{H}_A \otimes \hat{H}_B$, $\hat{U}_A \otimes \hat{U}_B$, and $\hat{\rho}_A \otimes \hat{\rho}_B$ are self-adjoint, unitary, and a density operator acting on $\mathcal{H}$, respectively. But keep in mind that this does not mean that all self-adjoint, unitary, or density operators acting on $\mathcal{H}$ can be written in a simple tensor product form $\hat{L}_A \otimes \hat{L}_B$.

1.3 The quantum-mechanical framework

In this section we review the basic postulates that describe how quantum mechanics treats physical systems. As the building blocks of the theory, these axioms cannot be *proved*. They can only be formulated following *plausibility arguments* based on the *observation* of physical phenomena and the *connection* of the theory with previous theories which are known to work in some limit. We will try to motivate (and justify to a point) these axioms as much as possible, starting with a brief historical introduction to the context in which they were created[15].

1.3.1 A brief historical introduction

By the end of the 19th century there was a great feeling of safety and confidence among the physics community: analytical mechanics (together with statistical mechanics) and Maxwell's electromagnetism (in the following *classical physics* altogether) seem to explain the whole range of physical phenomena that one could observe, and hence, in a sense, the foundations of physics were complete. There were, however, a couple of experimental observations which lacked explanation

---

[15] For a thorough historical overview of the birth of quantum physics see [9].

within this 'definitive' framework, which actually led to the construction of a whole new way of understanding physical phenomena: quantum mechanics.

Among this experimental evidence, the shape of the high-energy spectrum of the radiation emitted by a black body, the photoelectric effect which showed that only light exceeding some frequency can release electrons from a metal irrespective of its intensity, and the discrete set of spectral lines of hydrogen, were the principal triggers of the revolution to come in the first quarter of the 20th century. The first two led Planck and Einstein to suggest that electromagnetic energy is not continuous but divided into small packets of energy $\hbar\omega$ (ω being the angular frequency of the radiation), while Bohr succeeded in explaining the hydrogen spectrum by assuming that the electron orbiting the nucleus can occupy only a discrete set of orbits with angular momenta proportional to $\hbar$. The constant $\hbar = h/2\pi \sim 10^{-34}$ J $\cdot$ s, where h is now known as the Planck constant, appeared in both cases as the 'quantization unit', the value separating the quantized values that energy or angular momentum are able to take.

Even though the physicists of the time tried to understand this quantization of the physical magnitudes within the framework of classical physics, it was soon realized that a completely new theory was required. The first attempts to build such a theory (which actually worked for some particular scenarios) were based on applying ad hoc quantization rules to various mechanical variables of systems, but with a complete lack of physical interpretation for such rules [10]. However, between 1925 and 1927 the first real formulations of the necessary theory were developed: the *wave mechanics* of Schrödinger [11] and the *matrix mechanics* of Heisenberg, Born and Jordan [12–14] (see [10] for English translations), which also received independent contributions from Dirac [15]. Even though in both theories the quantization of various observable quantities appeared naturally and in correspondence with experiments, they seemed completely different, at least until Schrödinger showed the equivalence between them.

The new theory was later formalized mathematically using vector spaces by Dirac [5] (although not entirely rigorously), and a little later by von Neumann with more mathematical rigor using Hilbert spaces [16] (see [17] for an English version). They developed the laws of *quantum mechanics* basically as we know them today [18–23]. In the following sections we will introduce these rules in the form of six axioms that will set out the structure of the theory of quantum mechanics as will be used throughout this book.

1.3.2 Axiom 1. Observables and measurement outcomes

The experimental evidence for the tendency of observable physical quantities to be quantized at the microscopic level motivates the first axiom:

> **Axiom 1.** Any physical observable quantity A corresponds to a self-adjoint operator $\hat{A}$ acting on an abstract Hilbert space. After a measurement of A, the only possible outcomes are the eigenvalues of $\hat{A}$.

The quantization of physical observables is therefore directly introduced within the theory by this postulate. Note that it does not say anything about the dimension d of the Hilbert space corresponding to a given observable, and it even leaves open the possibility of observables having a continuous spectrum, rather than a discrete one. The problem of how to make the proper correspondence between observables and self-adjoint operators will be addressed in an axiom to come.

In this book we use the name 'observable' both for the physical quantity A and its associated self-adjoint operator $\hat{A}$ indistinctly. Observables having purely discrete or purely continuous spectra will be referred to as *countable* and *continuous observables*, respectively.

1.3.3 Axiom 2. The state of the system and statistics of measurements

The next axiom follows from the following question: according to the previous axiom the eigenvalues of an observable are the only values that can appear when measuring it, but what about the statistics of such a measurement? We know a class of operators in Hilbert spaces which act as probability distributions for the eigenvalues of any self-adjoint operator, density operators. This motivates the second axiom:

Axiom 2. The state of the system is completely specified by a density operator $\hat{\rho}$. When measuring a countable observable A with eigenvectors $\{|a_j\rangle\}_{j=1,2,...,d}$ (d might be infinite), associated with the possible outcomes $\{a_j\}_{j=1,2,...,d}$ is a probability distribution $\{p_j = \rho_{jj}\}_{j=1,2,...,d}$ which determines the statistics of the experiment (the *Born rule*). Similarly, when measuring a continuous observable X with eigenvectors $\{|x\rangle\}_{x\in\mathbb{R}}$, the probability density function $P(x) = \rho(x, x)$ is associated with the possible outcomes $\{x\}_{x\in\mathbb{R}}$ in the experiment.

This postulate has deep consequences that we analyze now. Contrary to classical mechanics (and intuition), even if the system is prepared in a given state, the value of an observable is in general not well defined. We can only specify with what probability a given value of the observable will come out in a measurement. Hence, this axiom proposes a change of paradigm; determinism must be abandoned: the theory is no longer able to predict with certainty the outcome of a single run of an experiment in which an observable is measured, but rather gives the statistics that will be extracted after a large number of runs.

To be fair, there is a case in which the theory allows us to predict the outcome of the measurement of an observable with certainty: when the system is prepared such that its state is an eigenvector of the observable. This seems much like when in classical mechanics the system is prepared with a given value of its observables. However, we will show that it is impossible to find a common eigenvector to *all* the available observables of a system, and hence the difference between classical and quantum mechanics is that in the latter it is impossible to prepare the system in a state which would allow us to predict with certainty the outcome of a measurement of each of its observables. Let us try to elaborate on this in a more rigorous fashion.

Let us define the *expectation value* of a given operator $\hat{B}$ as

$$\langle \hat{B} \rangle = \text{tr}\{\hat{\rho}\hat{B}\}. \tag{1.23}$$

In the case of a countable observable $\hat{A}$ or a continuous observable $\hat{X}$, this expectation value can be written in their own eigenbases as

$$\langle \hat{A} \rangle = \sum_{j=1}^{d} \rho_{jj} a_j \qquad \text{and} \qquad \langle \hat{X} \rangle = \int_{-\infty}^{+\infty} \text{d}x \rho(x, x)x, \tag{1.24}$$

which correspond to the mean value of the outcomes registered in a large number of measurements of the observables. We define also the *variance* $V(A)$ of the observable as the expectation value of the square of its *fluctuation operator* $\delta\hat{A} = \hat{A} - \langle \hat{A} \rangle$, that is,

$$V(A) = \text{tr}\left\{\hat{\rho}\left(\delta\hat{A}\right)^2\right\} = \langle \hat{A}^2 \rangle - \langle \hat{A} \rangle^2, \tag{1.25}$$

from which we obtain the *standard deviation* or *uncertainty* as $\Delta A = \sqrt{V(A)}$, which measures how much the outcomes of the experiment deviate from the mean, and hence, somehow specifies how 'well defined' the value of the observable A is.

Note that the probability of obtaining the outcome a_j when measuring A can be written as the expectation value of the projection operator $\hat{P}_j = |a_j\rangle\langle a_j|$, that is $p_j = \langle \hat{P}_j \rangle$. Similarly, the probability density function associated with the possible outcomes $\{x\}_{x\in\mathbb{R}}$ when measuring X can be written as $P(x) = \langle \hat{P}(x) \rangle$, where $\hat{P}(x) = |x\rangle\langle x|$.

Having written all these objects (probabilities, expectation values, and variances) in terms of traces is really useful, since the trace is invariant under basis changes, and hence can be evaluated in any basis we want to work with, see section 1.2.2.

These axioms have one further counterintuitive consequence. It is possible to prove that irrespectively of the state of the system, the following relation between the variances of two non-commuting observables A and B is satisfied:

$$\Delta A \Delta B \geq \frac{1}{2}\left|\langle [\hat{A}, \hat{B}] \rangle\right|. \tag{1.26}$$

According to this inequality, known as the *uncertainty principle* (which was first derived by Heisenberg), in general, the only way in which the observable A can be perfectly defined ($\Delta A \to 0$) is by making observable B completely undefined ($\Delta B \to \infty$), or vice versa. Hence, in the quantum formalism one cannot, in general, prepare the system in a state in which all its observables are well defined, the complete opposite to our everyday experience.

Before moving to the third axiom, let us comment on a couple more things related to the state of the system. It is possible to show that a density operator can always be expressed as a *statistical* or *convex mixture* of projection operators, that is, $\hat{\rho} = \sum_{m=1}^{M} w_m |\varphi_m\rangle\langle\varphi_m|$, where $\{w_m\}_{m=1,2,...,M}$ is a probability distribution and the vectors $\{|\varphi_m\rangle\}_{m=1}^{M}$ are normalized to one, but do not need to be orthogonal (note that in fact M does not need to be equal to d). Hence, another way of specifying the state of the system is by a set of normalized vectors together with some statistical rule for mixing them, that is, the set $\{w_m, |\varphi_m\rangle\}_{m=1,2,...,M}$, known as an *ensemble decomposition* of the state $\hat{\rho}$. Such decompositions are not unique, in the sense that different ensembles can lead to the same $\hat{\rho}$. It can be proved though [24] that two ensembles $\{w_m, |\varphi_m\rangle\}_{m=1,2,...,M}$ and $\{v_n, |\psi_n\rangle\}_{n=1,2,...,N}$ (we take $M \leqslant N$ for definiteness) give rise to the same density operator $\hat{\rho}$ if and only if there exists a left-unitary matrix[16] U with elements $\{U_{mn}\}_{m,n=1,2,...,N}$ such that [24]

$$\sqrt{w_m}|\varphi_m\rangle = \sum_{n=1}^{N} U_{mn}\sqrt{v_n}|\psi_n\rangle, \quad m = 1, 2, ..., N, \tag{1.27}$$

where if $M \neq N$, $N - M$ zeros must be included in the ensemble with fewer states, so that $\mathcal{U}$ is a square matrix.

When only one vector $|\varphi\rangle$ contributes to the mixture, $\hat{\rho} = |\varphi\rangle\langle\varphi|$ is completely specified by just this single vector, and we say that the density operator is *pure*; otherwise, we say that it is *mixed*. A necessary and sufficient condition for $\hat{\rho}$ to be pure is $\hat{\rho}^2 = \hat{\rho}$. In the next chapters we will learn that the mixedness of a state always comes from the fact that some of the information of the system has been lost to some other inaccessible system with which it has interacted for a while before becoming isolated. In other words, the state of a system is pure only when it has no correlations at all with other systems.

Note, finally, that when the state of the system is in a *pure state* $|\psi\rangle$, the expectation value of an operator $\hat{B}$ takes the simple form $\langle\psi|\hat{B}|\psi\rangle$. Moreover, the pure state can be expanded in the countable and continuous bases of two observables $\hat{A}$ and $\hat{X}$ as

$$|\psi\rangle = \sum_{j=1}^{d} \psi_j |a_j\rangle \qquad \text{and} \qquad |\psi\rangle = \int_{-\infty}^{+\infty} dx \psi(x)|x\rangle, \tag{1.28}$$

respectively, being $\psi_j = \langle a_j|\psi\rangle$ and $\psi(x) = \langle x|\psi\rangle$. In this case, the probability distribution for the discrete outcomes $\{a_j\}_{j=1,2,...,d}$ is given by $\{p_j = |\psi_j|^2\}_{j=1,2,...,d}$, while the probability density function for the continuous outcomes $\{x\}_{x\in\mathbb{R}}$ is given by $P(x) = |\psi(x)|^2$.

1.3.4 Axiom 3. Composite systems

The next axiom specifies how the theory accommodates dealing with composite systems within its mathematical framework. Of course, a composition of two

---

[16] U is left-unitary if $U^\dagger U = I$ but $UU^\dagger$ might not be I, where I is the identity matrix of the corresponding dimension. It is easy to prove that finite-dimensional left-unitary matrices are unitary.

systems is itself another system subject to the laws of quantum mechanics; the question is how can we construct it.

Axiom 3. Consider two systems A and B with associated Hilbert spaces $\mathcal{H}_A$ and $\mathcal{H}_B$. Then, the state of the composite system as well as its observables act onto the tensor product Hilbert space $\mathcal{H}_{AB} = \mathcal{H}_A \otimes \mathcal{H}_B$.

This axiom has the following consequence. Imagine that the systems A and B interact during some time in such a way that they can no longer be described by independent states $\hat{\rho}_A$ and $\hat{\rho}_B$ acting on $\mathcal{H}_A$ and $\mathcal{H}_B$, respectively, but by a state $\hat{\rho}_{AB}$ acting on the joint space $\mathcal{H}_{AB}$. After the interaction, system B is kept isolated from any other system, but system A is given to an observer, who is therefore able to measure observables defined in $\mathcal{H}_A$ only, and might not even know that system A is part of a larger system. The question is, is it possible to reproduce the statistics of the measurements performed on system A with some state $\hat{\rho}_A$ acting on $\mathcal{H}_A$ only? This question has a positive and *unique* answer: this state is given by the *reduced density operator* $\hat{\rho}_A = \mathrm{tr}_B\{\hat{\rho}_{AB}\}$, that is, by performing the partial trace[17] with respect to system B's subspace onto the joint state.

1.3.5 Axiom 4. Quantization rules

The introduction of the fourth axiom is motivated by the following fact. The class of self-adjoint operators forms a real vector space with respect to the addition of operators and the multiplication of an operator by a real number. Using the commutator we can also build an operation that takes two self-adjoint operators $\hat{A}$ and $\hat{B}$ to generate another self-adjoint operator $\hat{C} = \mathrm{i}[\hat{A}, \hat{B}]$, which, in addition, satisfies all the properties required by a Lie product. Hence, even if classical and quantum theories seem fundamentally different, it seems that observables are treated similarly within their corresponding mathematical frameworks: they are elements of a Lie algebra.

On the other hand, we saw that the generalized coordinates and momenta have a particular mathematical structure in the Hamiltonian formalism, they are the generators of the Heisenberg group. It seems then quite reasonable to ask for the same in the quantum theory, so that at least concerning observables both theories are equivalent. This motivates the fourth axiom:

---

[17] Given an orthonormal basis $\{|b_j\rangle\}_j$ of $\mathcal{H}_B$, this is defined by

$$\mathrm{tr}_B\{\hat{\rho}_{AB}\} = \sum_j \langle b_j|\hat{\rho}_{AB}|b_j\rangle, \tag{1.29}$$

which is indeed an operator acting on $\mathcal{H}_A$.

Axiom 4. Consider a physical system which is described classically within a Hamiltonian formalism by a set of generalized coordinates $\mathbf{q} = \{q_j\}_{j=1}^n$ and momenta $\mathbf{p} = \{p_j\}_{j=1}^n$ at a given time. Within the quantum formalism, the corresponding observables $\hat{\mathbf{q}} = \{\hat{q}_j\}_{j=1}^n$ and $\hat{\mathbf{p}} = \{\hat{p}_j\}_{j=1}^n$ satisfy the *canonical commutation relations*

$$\left[\hat{q}_j, \hat{p}_l\right] = i\hbar\delta_{jl} \quad \text{and} \quad \left[\hat{q}_j, \hat{q}_l\right] = \left[\hat{p}_j, \hat{p}_l\right] = 0. \tag{1.30}$$

The constant $\hbar$ is included because, while the Poisson bracket $\{q_j, p_l\}$ has no units, the commutator $[\hat{q}_j, \hat{p}_l]$ has units of action. That it is exactly $\hbar$ the proper constant can be seen only once the theory is compared with experiments.

We can now discuss how to build the self-adjoint operator corresponding to a given observable. In general, meaningful observables are built from symmetry principles [23], e.g. the kinetic and angular momenta as the generators of space translations and rotations, respectively. An alternative route might be taken when the observable is well-defined classically. Suppose that in the Hamiltonian formalism the observable A is represented by the real phase-space function $A(\mathbf{q}, \mathbf{p})$. It seems quite natural to use then $A(\hat{\mathbf{q}}, \hat{\mathbf{p}})$ as the corresponding quantum operator. However, this correspondence faces a lot of troubles resulting from the fact that, while coordinates and momenta commute in classical mechanics, they do not in quantum mechanics. For example, given the classical observable $A = qp = pq$, we could be tempted to assign to it any of the quantum operators $\hat{A}_1 = \hat{q}\hat{p}$ or $\hat{A}_2 = \hat{p}\hat{q}$. These two operators are not equivalent (they do not commute) and they are not even self-adjoint, and hence cannot represent observables. One possible solution to this problem, at least for observables with a series expansion, is to always symmetrize the classical expressions with respect to coordinates and momenta, so that the resulting operator is self-adjoint. Applied to our previous example, we should take $\hat{A} = (\hat{p}\hat{q} + \hat{q}\hat{p})/2$ according to this rule. This simple procedure leads to the correct results most of the times, and when it fails (for example, if the classical observable does not have a series expansion) it was proved by Groenewold [25] that it is possible to make a faithful systematic correspondence between classical observables and self-adjoint operators by using more sophisticated correspondence rules.

Of course, when the observable corresponds to a degree of freedom which is not defined in a classical context (such as *spin*), it must be built from scratch based on experimental observations and/or first principles.

Note that the commutation relations between coordinates and momenta makes them satisfy the uncertainty relation $\Delta q \Delta p \geqslant \hbar/2$, and hence, if one of them is well defined in the system, the other must have statistics very spread around the mean.

1.3.6 Axiom 5. *Free* evolution of the system

The previous axioms have served to define the mathematical structure of the theory and its relation to physical systems. We have not said anything yet about how quantum mechanics treats the evolution of the system. As we are about to see, the formalism treats very differently the evolution due to a measurement performed by an observer, and the *free* evolution of the system when it is not subject to observation. The following axiom specifies how to deal with the latter case. Just as with the previous axiom, it feels pretty reasonable to keep the analogy with the Hamiltonian formalism, a motivation which comes also from the fact that, as stated, quantum mechanics must converge to classical mechanics in some limit. In the Hamiltonian formalism, observables evolve according to (1.7), so that making the correspondence between the classical and quantum Lie products as in the previous axiom, we enunciate the fifth axiom:

Axiom 5. The evolution of an observable $\hat{A}(\hat{q}, \hat{p}, ...; t)$ is given by

$$i\hbar \frac{d\hat{A}}{dt} = \left[\hat{A}, \hat{H}\right] + \frac{\partial \hat{A}}{\partial t}, \tag{1.31}$$

---

which is known as the Heisenberg equation, and where $\hat{H}(\hat{q}, \hat{p},...;t)$ is the self-adjoint operator corresponding to the Hamiltonian of the system. Note that the notation '$\hat{q}$, $\hat{p}$, ...' emphasizes the fact that the observable may depend on fundamental operators other than the generalized coordinates, e.g. purely quantum degrees of freedom such as spin.

For the case of an observable and a Hamiltonian with no explicit time-dependence (as will be assumed from now on), this evolution equation admits the explicit solution

$$\hat{A}(t) = \hat{U}^{\dagger}(t)\hat{A}(0)\hat{U}(t), \quad \text{being } \hat{U}(t) = \exp\left[\hat{H}t/i\hbar\right], \tag{1.32}$$

a unitary operator called the *evolution operator*. For explicitly time-dependent Hamiltonians it is still possible to solve formally the Heisenberg equation as a *Dyson series*, but we will not worry about this case, as it does not appear throughout the monograph. Let us remark that this type of evolution ensures that if the canonical commutation relations (1.30) are satisfied at some time, they will be satisfied at all times.

Note that within this formalism the state $\hat{\rho}$ of the system is fixed in time, the observables are the ones which evolve. On the other hand, we have seen that, concerning observations (experiments), only expectation values of operators are relevant; and for an observable $\hat{A}$ at time t, these can be written as

$$\langle\hat{A}(t)\rangle = \text{tr}\left\{\hat{\rho}\hat{A}(t)\right\} = \text{tr}\left\{\hat{U}(t)\hat{\rho}\hat{U}^{\dagger}(t)\hat{A}(0)\right\}, \tag{1.33}$$

where in the last equality we have used the cyclic property of the trace. This expression shows that, instead of treating the observable as the evolving operator, we can define a new state at time t given by

$$\rho(t) = \hat{U}(t)\hat{\rho}(0)\hat{U}^{\dagger}(t), \tag{1.34}$$

while keeping fixed the operator. In differential form, this expression reads

$$i\hbar\frac{\mathrm{d}\hat{\rho}}{\mathrm{d}t} = \left[\hat{H}, \hat{\rho}\right], \tag{1.35}$$

which is known as the *von Neumann equation*. When the system is in a pure state $|\psi\rangle$, the following evolution equation is derived for the state vector itself

$$i\hbar\frac{\mathrm{d}}{\mathrm{d}t}|\psi\rangle = \hat{H}|\psi\rangle, \tag{1.36}$$

which is known as the *Schrödinger equation*, from which the state at time t is found as $|\psi(t)\rangle = \hat{U}(t)|\psi(0)\rangle$.

Therefore, we have two different but equivalent evolution formalisms or *pictures*. In one, which we will call the *Heisenberg picture*, the state of the system is fixed, while observables evolve according to the Heisenberg equation. In the other, which we will call the *Schrödinger picture*, observables are fixed, while states evolve according to the von Neumann equation.

1.3.7 Axiom 6. Post-measurement state

The previous postulate specifies how the free evolution of the system is taken into account in the quantum-mechanical formalism. It is then left to specify how the state of the system evolves after a measurement is performed on it. For reasons that we will briefly review after enunciating the axiom, this is probably the most controversial point in the quantum formalism. Indeed, while in classical physics we assume that it is possible to perform measurements onto the system without disturbing its state, this final quantum-mechanical axiom states:

Axiom 6. If upon a measurement of a countable observable A the outcome a_m is obtained, then immediately after the measurement the state of the system *collapses* to $|a_m\rangle$.

Before commenting on the controversial aspects of this axiom, it is important to mention some operational and conceptual aspects that will be important later. Let us denote by $\hat{\rho}$ and $\hat{\rho}_m$ the states before and after the measurement is performed. It is convenient to define the *unnormalized post-measurement state* $\tilde{\rho}_m = \hat{P}_m\hat{\rho}\hat{P}_m$, where we recall that $\hat{P}_m = |a_m\rangle\langle a_m|$ is a projector. The normalized post-measurement state can be obtained then as $\hat{\rho}_m = p_m^{-1}\tilde{\rho}_m$, where the probability of obtaining the a_m outcome can be evaluated as the trace of the unnormalized state, $p_m = \mathrm{tr}\{\tilde{\rho}_m\}$.

The axiom assumes that, after the measurement, the observer gains knowledge about the measurement outcome, which we will denote as a *selective* measurement. However, suppose that for some reason the user interface of the measurement device does not allow us to distinguish between a set of outcomes $\{a_{m_k}\}_{k=1,2,\ldots,K}$, which we will denote by a *partially selective* measurement. Then, after the corresponding experimental outcome is obtained, the best estimate that the observer can assign to the post-measurement state is the ensemble decomposition $\{\bar{p}_{m_k}, |a_{m_k}\rangle\}_{k=1,2,...,K}$ with relative probabilities $\bar{p}_{m_k} = p_{m_k}/\sum_{k=1}^{K} p_{m_k}$, since the real outcome is unknown, but the *a priori* probabilities p_m of the possible outcomes are known. Hence, in such case we would assign the post-measurement state $\hat{\rho}_{\{m_1,m_2,...,m_K\}} = \sum_{k=1}^{K} \bar{p}_{m_k}\hat{\rho}_{m_k}$ to the system. The extreme case in which the outcome of the measurement is simply not recorded, so that we cannot know which outcome occurred and the best estimate for the post-measurement state is $\hat{\rho}' = \sum_{m=1}^{d} p_m\hat{\rho}_m = \sum_{m=1}^{d} \tilde{\rho}_m$, is known as a *non-selective* measurement.

So far we have considered the post-measurement state in the case of measuring a countable observable. The continuous case is tricky, since, as mentioned in section 1.2.3, the eigenvectors of a continuous observable cannot correspond to physical states (they cannot be normalized). On the other hand, one can always argue that the detection of a single definite value out of the spectrum $\{x\}_{x\in\mathbb{R}}$ of a continuous observable $\hat{X}$ would require an infinite precision, whereas detectors always have some finite precision. Consequently, there are two natural ways of dealing with such a problem:

- Accepting that measuring continuous observables is simply not possible, and what is measured in real experiments is always some countable version of them, which only in some unphysical limit reproduce the continuous measurement precisely. An example of this consists in the process of *binning* the continuous observable, which assumes that the detector can only distinguish between pixels with width Δ_x centered at certain points $\{x_k = k\Delta_x\}_{k\in\mathbb{Z}}$ in the spectrum of the continuous observable, so what is measured is instead the countable observable

$$\hat{X}_{\text{count}} = \sum_{k=-\infty}^{\infty} x_k|x_k\rangle\langle x_k|, \quad \text{with } |x_k\rangle = \frac{1}{\sqrt{\Delta_x}} \int_{x_k-\Delta_x/2}^{x_k+\Delta_x/2} \mathrm{d}x|x\rangle. \quad (1.37)$$

- Allowing for the possible outcomes of the measurement to still be continuous, but with an uncertainty given by the precision of the measurement device. In the case of starting with a pure state, this would simply mean that the post-measurement state is not an eigenvector of the continuous operator, but a normalizable superposition of several of them, spanning around the measured value with a width given by the measurement's precision. This intuitive approach can be formalized with the theory of generalized quantum measurements that we will see later in this book [26, 27].

In any case, it is sometimes useful for theoretical calculations to proceed as if perfectly precise measurements were possible, with the system collapsing to one eigenvector of the continuous observable. However, it is important to keep in mind that this is just an unphysical idealization, whose corrections have to be taken into account when applying it to a real situation.

We can pass now to discuss the controversial aspects of this axiom, of which a pedagogical introduction can be found in [22] (see also [28] and appendix E of [7]).

In short, the problem is that, even though it leads to predictions which fully agree with the observations, the axiom somehow creates an *inconsistency* in the theory because of the following argument. According to axiom 5, the unitary evolution of a system not subject to observation is *reversible*, that is, one can always change the sign of the relevant terms of the Hamiltonian which contributed to the evolution (at least conceptually), and come back to the original state. On the other hand, the *collapse* axiom claims that when the system is put in contact with a measurement device and an observable is measured, the state of the system collapses to some other state in an *irreversible* way[18]. However, coming back to axiom 5, the whole measurement process could be described reversibly by considering, in addition to the system's particles, the evolution of all the particles forming the measurement device (or even the human who is observing the measurement outcome if needed!), that is, the Hamiltonian for the whole 'observed system + measurement device' scenario. Hence, it seems that, when including the collapse axiom, quantum mechanics allows for two completely different descriptions of the measurement process, one reversible and one irreversible, without giving a clear rule for when to apply each. It is in this sense that the theory contains an inconsistency.

There are three[19] main positions that physicists have taken regarding how this inconsistency might be solved, which we will refer too as *objective*, *subjective*, and *apparent* collapse interpretations, and whose (highly simplified) main ideas we discuss here[20]:

- **Objective collapse.** There is a clear boundary (yet to be found) between the quantum and the classical worlds. In the classical world, to which measurement devices and observers belong, there exists some *decoherence mechanism* that prevents systems from being in a superposition of states corresponding to mutually exclusive values of their observables. When the measurement device enters in contact with the quantum system, the latter becomes a part of the classical world, and the aforementioned decoherence mechanism forces its collapse. Hence, within this interpretation the collapse is pretty much a real physical process that we still need to understand along with the quantum/classical boundary. There are several *collapse theories* available at the moment [28], some of which we expect to be able to falsify or confirm in the near future with modern quantum technologies based on, for example, opto-, electro-, or magneto-mechanics [31].

---

[18] Note that in the literature the terms *reversible* and *irreversible* are sometimes replaced by *linear* and *non-linear*, referring to the fact that unitary evolution comes from a linear equation (Schrödinger or von Neumann equations), while measurement-induced dynamics becomes non-linear through its dependence on the probability of the possible outcomes, which in turn depend on the state.

[19] But each with many sub-interpretations differing in subtle, or even not so subtle points. In essence, one can find a lot of truth to the saying 'Give me a room with N physicists and I'll find you $N + 1$ different interpretations of quantum mechanics'.

[20] One interpretation of quantum mechanics that is formulated with a completely different set of axioms, and hence does not fit this list is *Bohmian mechanics* [29, 30], in which particles follow deterministic trajectories, but determined from a *guiding wave* that obeys the Schrödinger equation.

- **Subjective collapse.** The state is simply a mathematical object which conveniently describes the statistics of experiments, but that otherwise has no physical significance. As such, what the quantum formalism provides is simply a set of rules for how to update our best estimate to the state according to the information that we have about the system. In this sense, the collapse is just the way that an observer subjectively updates the state of the system after gathering the information concerning the outcome. Quantum Bayesianism or *QBism* [32, 33] is possibly the most refined of such interpretations, and has gathered a lot of momentum in recent years.
- **Apparent collapse.** The measurement can be described without the need to abandon the framework of axiom 5 as a joint unitary transformation onto the system and the measurement device (even including the observer), leading to a final entangled state[21] between those in which the eigenstates of the system's observable are in one-to-one correspondence with a set of macroscopically distinct *pointer* states of the measurement device [22]. Hence, after a measurement, reality splits into many *branches* where observers experience different outcomes and which stay in a quantum superposition, and collapse appears just an illusion coming from the fact that we only see the effective dynamics projected into the corresponding branch that we are experiencing. This approach finds its best-developed expression in the so-called *many-worlds interpretation* [34], which describes the quantum-mechanical framework as unitary evolution of a pure state (*wave function*) of the whole Universe, which includes all the branches or *worlds*[22].

In any case, whether of objective, subjective, or apparent value, it is clear that the collapse axiom is of great *operational* value, that is, it is currently the easiest successful way of analyzing schemes involving measurements, and hence we will apply it when needed.

1.3.8 The von Neumann entropy

An object of fundamental relevance in the theory of quantum information is the *von Neumann entropy* of a state $\hat{\rho}$, which is defined as

$$S[\hat{\rho}] = -\text{tr}\{\hat{\rho} \log \hat{\rho}\}. \tag{1.38}$$

Given the diagonal representation of the state

$$\hat{\rho} = \sum_{j=1}^{d} \lambda_j |r_j\rangle\langle r_j|, \tag{1.39}$$

---

[21] We will discuss the concept of entanglement in detail in the next chapter.

[22] Believers of the many-worlds interpretation, or even people who are not sure of their quantum-mechanical beliefs (better safe than sorry), are strongly encouraged to use the *quantum world splitter* when they need to choose between equally reasonable options in life: www.qol.tau.ac.il.

where its eigenvectors $\{|r_j\rangle\}_{j=1,2,\ldots,d}$ form an orthonormal basis of the Hilbert space $\mathcal{H}$ (which therefore has dimension d), the von Neumann entropy is just the Shannon entropy of the distribution[23] $\lambda = \mathrm{col}(\lambda_1, \lambda_2, \ldots, \lambda_d)$, that is,

$$S[\hat{\rho}] = -\sum_{j=1}^{d} \lambda_j \log \lambda_j. \qquad (1.40)$$

For a pure state $|\varphi\rangle\langle\varphi|$ the entropy is zero, while it has a maximum $\log d$ for the *maximally mixed state*[24]

$$\hat{\rho}_{\mathrm{MM}} = \frac{1}{d}\hat{I}. \qquad (1.41)$$

This suggests that the von Neumann entropy can be understood as a measure of the mixedness of the state. Note that when the state of the system is $\hat{\rho}_{\mathrm{MM}}$ and some observable $\hat{A}$ is measured, all its eigenvalues are equally likely to appear as an outcome of the measurement, that is,

$$p_j = \langle a_j|\hat{\rho}_{\mathrm{MM}}|a_j\rangle = \frac{1}{d} \quad \forall j. \quad \text{(flat distribution)} \qquad (1.42)$$

This reinforces the interpretation of mixedness as due to some kind of information loss.

The entropy is a strictly concave functional of density operators, that is, given the convex mixture of states

$$\hat{\rho} = \sum_{m=1}^{M} w_m \hat{\rho}_m, \qquad (1.43)$$

where $\{w_m\}_{m=1,2,\ldots,M}$ is a probability distribution and $\{\hat{\rho}_m\}_{m=1,2,\ldots,M}$ is a set of density operators, we have [24]

$$S[\hat{\rho}] \geqslant \sum_{m=1}^{M} w_m S[\hat{\rho}_m], \qquad (1.44)$$

where the equality holds only when all the states ρ_m with non-zero probability w_m are equal. This property agrees very well with the interpretation of mixedness as coming from information loss: the average of the information contained in each state $\hat{\rho}_m$ cannot be smaller than the information contained in the state $\hat{\rho}$ obtained by scrambling them according to the probability distribution $\mathbf{w}$, since the latter imposes an extra level of randomness.

It is interesting to note (and easy to prove mathematically [24]) that the entropy does not change by unitary evolution, can only increase by non-selective projective

---

[23] In the following we use the notations $\mathbf{p} = \mathrm{col}(p_1, p_2,\ldots, p_d)$ and $\{p_j\}_{j=1,2,\ldots,d}$ for a probability distribution interchangeably. Remember that 'col' is short for 'column vector'.

[24] The case $d \to \infty$ will be discussed later when studying infinite-dimensional spaces in detail in the last chapter.

measurements, and can only decrease by selective projective measurements (indeed, when no degeneracies are present, it collapses to zero, as the state becomes pure). These properties are indeed expected from a purely informational point of view: while evolving 'freely' the system does not exchange any information with any other system; when we perform a selective measurement indeed we gain information about the system and its post-measurement state; when the measurement is non-selective, information can become 'lost' in the measurement device, but we definitely can never obtain more information about the system than we already had prior to the measurement.

Bibliography

[1] Goldstein H, Poole C and Safko J 2001 *Classical Mechanics* (Reading, MA: Addison-Wesley)
[2] Greiner W 1989 *Classical Mechanics: Point Particles and Relativity* (Berlin: Springer)
[3] Greiner W and Reinhardt J 1989 *Classical Mechanics: Systems of Particles and Hamiltonian Dynamics* (Berlin: Springer)
[4] Hand L N and Finch J D 1998 *Analytical Mechanics* (Cambridge: Cambridge University Press)
[5] Dirac P A M 1930 *The Principles of Quantum Mechanics* (Oxford: Oxford University Press)
[6] Prugovečky E 1971 *Quantum Mechanics in Hilbert Space* (New York: Academic)
[7] Galindo A and Pascual P 1990 *Quantum Mechanics* vol 1 (Berlin: Springer)
[8] Gelfand I M and Vilenkin N Y 1964 *Generalized Functions* vol 4 (New York: Academic)
[9] Whitaker A 1996 *Einstein, Bohr, and the Quantum Dilemma* (Cambridge: Cambridge University Press)
[10] van der Waerden B L 1968 *Sources of Quantum Mechanics* (New York: Dover)
[11] Schrödinger E 1926 An undulatory theory of the mechanics of atoms and molecules *Phys. Rev.* **28** 1049
[12] Heisenberg W 1925 Über quantentheoretische Umdeutung kinematischer und mechanischer Beziehungen *Z. Phys.* **33** 879
[13] Born M and Jordan P 1925 Zur Quantenmechanik *Z. Phys.* **34** 858
[14] Born M, Heisenberg W and Jordan P 1926 Zur Quantenmechanik II *Z. Phys.* **35** 557
[15] Dirac P A M 1926 The fundamental equations of quantum mechanics *Proc. R. Soc.* A **109** 642
[16] von Neumann J 1932 *Mathematische Grundlagen der Quantenmechanik* (Berlin: Springer)
[17] von Neumann J 1955 *Mathematical Foundations of Quantum Mechanics* (Princeton, NJ: Princeton University Press)
[18] Cohen-Tannoudji C, Diu B and Laloë F 1977 *Quantum Mechanics* vol 1 (New York: Wiley)
[19] Cohen-Tannoudji C, Diu B and Laloë F 1977 *Quantum Mechanics* vol 2 (New York: Wiley)
[20] Greiner W and Müller B 1989 *Quantum Mechanics: An Introduction* (Berlin: Springer)
[21] Greiner W 1989 *Quantum Mechanics: Symmetries* (Berlin: Springer)
[22] Basdevant J-L and Dalibard J 2002 *Quantum Mechanics* (Berlin: Springer)
[23] Ballentine L E 1998 *Quantum Mechanics: A Modern Development* (Singapore: World Scientific)
[24] Nielsen M A and Chuang I L 2000 *Quantum Information and Quantum Computation* (Cambridge: Cambridge University Press)
[25] Groenewold H J 1946 On the principles of elementary quantum mechanics *Physica* **12** 405
[26] Wiseman H M and Milburn G J 2009 *Quantum Measurement and Control* (Cambridge: Cambridge University Press)

[27] Jacobs K and Steck D A 2006 A straightforward introduction to continuous quantum measurement *Contemp. Phys.* **47** 279

[28] Bassi A and Ghirardi G C 2003 Dynamical reduction models *Phys. Rep.* **379** 257

[29] Holland P R 1993 *The Quantum Theory of Motion* (Cambridge: Cambridge University Press)

[30] Oriols X and Mompart J 2012 *Applied Bohmian Mechanics: From Nanoscale Systems to Cosmology* (Singapore: Pan Stanford Publishing)

[31] Romero-Isart O 2011 Quantum superposition of massive objects and collapse models *Phys. Rev.* A **84** 052121

[32] Fuchs C A, Mermin N D and Schack R 2014 An introduction to QBism with an application to the locality of quantum mechanics *Am. J. Phys.* **82** 749

[33] Fuchs C A 2012 Interview with a quantum Bayesian arXiv: 1207.2141

[34] Vaidman L 2014 Many-worlds interpretation of quantum mechanics *The Stanford Encyclopedia of Philosophy* (Stanford, CA: Stanford University) http://plato.stanford.edu/entries/qm-manyworlds

An Introduction to the Formalism of Quantum Information
with Continuous Variables

Carlos Navarrete-Benlloch

Chapter 2

Bipartite systems and entanglement

In this chapter we will introduce some general concepts of entanglement theory, of which we will learn much more in section 4.4.4 when applying them to continuous-variable systems (including the origins of the theory, which can be traced back to the seminal work of Einstein, Podolsky, and Rosen [1]). For a more exhaustive and detailed introduction to the world of entanglement, see [2–4].

2.1 Entangled states

Consider two systems A and B (named after Alice and Bob, two observers who are able to interact locally with their respective system) with associated Hilbert spaces $\mathcal{H}_A$ and $\mathcal{H}_B$ of dimensions d_A and d_B, respectively. The systems are prepared in some state $\hat{\rho}_{AB}$ acting on the joint space $\mathcal{H}_A \otimes \mathcal{H}_B$. Recall from the discussion after axiom 3 that Alice and Bob can reproduce the statistics of measurements performed on their subsystems via the reduced states $\hat{\rho}_A = \mathrm{tr}_B\{\hat{\rho}_{AB}\}$ and $\hat{\rho}_B = \mathrm{tr}_A\{\hat{\rho}_{AB}\}$, respectively.

When the state of the joint system is of the type $\hat{\rho}_{AB}^{(\mathrm{prod})} = \hat{\rho}_A \otimes \hat{\rho}_B$, that is, a tensor product of two arbitrary density operators, the actions performed by Alice on system A will not affect Bob's system, the statistics of which are given by $\hat{\rho}_B$, no matter what the actual state $\hat{\rho}_A$ is. In this case A and B are completely *uncorrelated*. For any other type of joint state, A and B will share some kind of correlation.

Correlations appear naturally also within the classical framework. Hence, a problem of paramount relevance in quantum information is understanding which types of correlations can appear at a classical level, and which are purely quantum. This is because only if the latter are present, can one expect to use the correlated systems for quantum-mechanical applications which go beyond what is classically possible, the paradigmatic example being the exponential speed up of computational algorithms.

doi:10.1088/978-1-6817-4405-6ch2

Intuitively, the state of the system will only induce classical correlations between A and B when it can be written in the form

$$\hat{\rho}_{AB}^{(\text{sep})} = \sum_{m=1}^{M} w_m \hat{\rho}_A^{(m)} \otimes \hat{\rho}_B^{(m)}, \qquad (2.1)$$

where the $\hat{\rho}^{(m)}$ are density operators, $\{w_m\}_{k=1,2,...,M}$ is a probability distribution, M can be infinite, and the index m can even be continuous in some range, in which case the sum turns into an integral in that range and the probability distribution into a probability density function. Indeed, a state of the type (2.1) can be prepared by a protocol involving only local actions and classical correlations: Alice and Bob can share a classical machine which randomly picks a value of m according to the distribution $\{w_m\}_{m=1,2,...,M}$, and use it to trigger the preparation of the states $\hat{\rho}_A^{(m)}$ and $\hat{\rho}_B^{(m)}$, which can be done locally, and hence cannot induce further correlations. If the process is automatized so that Alice and Bob do not learn the outcome m of the random number generator, the best estimate that they can assign to the state is the mixture $\hat{\rho}_{AB}^{(\text{sep})}$. In other words, the state does not contain quantum correlations if it can be prepared using only *local operations and classical communication* (LOCC)[1].

There is yet another intuitive way of justifying that states which cannot be written in the separable form (2.1) will make A and B share quantum correlations. The idea is based on what is probably the most striking difference between classical and quantum mechanics: the *superposition principle*, that is, the possibility of states corresponding to *mutually exclusive* properties of the system to *interfere* (e.g. two different colors, $|\text{blue}\rangle + |\text{red}\rangle$). Hence, it is intuitive that correlations should have a quantum nature only when they come from some kind of superposition of joint states corresponding to mutually exclusive properties of the correlated systems (e.g. color in one system and flavor in the other, $|\text{blue}\rangle \otimes |\text{sweet}\rangle + |\text{red}\rangle \otimes |\text{sour}\rangle$), in which case the state cannot be written as a tensor product of two independent states ($|\text{blue}\rangle \otimes |\text{sour}\rangle$) or as a purely classical statistical mixture of these ($|\text{blue}\rangle\langle\text{blue}| \otimes |\text{sweet}\rangle\langle\text{sweet}| + |\text{red}\rangle\langle\text{red}| \otimes |\text{sour}\rangle\langle\text{sour}|$).

States of the type $\hat{\rho}_{AB}^{(\text{sep})}$ are called *separable*. Any *inseparable* state will induce quantum correlations between A and B. These correlations which cannot be generated by classical means are known as *entanglement*, and states which are not separable are called *entangled states*.

Given a bipartite state $\hat{\rho}_{AB}$, with corresponding reduced states $\hat{\rho}_A$ and $\hat{\rho}_B$, it is interesting to note that [5]

$$S(\hat{\rho}_{AB}) \leqslant S(\hat{\rho}_A) + S(\hat{\rho}_B), \qquad (2.2)$$

with the equality holding only when the systems A and B are uncorrelated, that is, $\hat{\rho}_{AB} = \hat{\rho}_A \otimes \hat{\rho}_B$. This property is known as the *subadditivity* of the von Neumann entropy, and clearly agrees with intuitive arguments based on information loss: the information contained in the state of the whole system cannot be obtained from the

---

[1] We will give a more precise meaning for this class of operations in the next chapter.

information left in the reduced states, since we have lost the information present in the correlations, which could not even be judged as being either classical or quantum merely from the reduced states.

2.2 Characterizing and quantifying entanglement

In general, given a mixed state $\hat{\rho}_{AB}$ acting on $\mathcal{H}_A \otimes \mathcal{H}_B$, it is hard to find out whether it is separable or not, the difficulty coming from distinguishing between quantum and classical correlations. Indeed, the best known criterion for separability, the *Peres–Horodecki criterion* [6, 7], yields only necessary and sufficient conditions when $d_A \times d_B \leqslant 6$ (note that this includes the case of two qubits), and for a reduced class of states in infinite-dimensional Hilbert spaces (some Gaussian states, see chapter 4). This criterion states that a necessary condition for the separability of a density operator is that it remains positive after the operation of partial transposition, that is, given

$$\hat{\rho}_{AB} = \sum_{jk=1}^{d_A} \sum_{lm=1}^{d_B} \rho_{jl,km} |a_j, b_l\rangle\langle a_k, b_m|, \tag{2.3}$$

where $\{|a_j, b_l\rangle = |a_j\rangle \otimes |b_l\rangle\}_{l=1,2,\ldots,d_B}^{j=1,2,\ldots,d_A}$ is an orthonormal basis of $\mathcal{H}_A \otimes \mathcal{H}_B$,

$$\hat{\rho}_{AB}^{T_B} = \sum_{jk=1}^{d_A} \sum_{lm=1}^{d_B} \rho_{jm,kl} |a_j, b_l\rangle\langle a_k, b_m|, \tag{2.4}$$

is a positive operator. We will learn more about this criterion and some more when studying infinite-dimensional Hilbert spaces in chapter 4.

A very different problem is that of quantifying the level of correlations present in the state, and more importantly, how much of these correspond to entanglement. Even though we understand fairly well the conditions that a proper *entanglement measure* $E[\hat{\rho}_{AB}]$ must satisfy, we have not found a completely satisfactory one for general states [2, 3] (either they do not satisfy all the conditions, or/and can only be efficiently computed for restricted classes of states). It is not the intention of this book to introduce in detail all these measures and explain up to what point they are satisfactory, but it is interesting to spend a few lines thinking about this issue, as it will allow us to obtain a better picture of what entanglement means (see [2, 3] for more details).

The basic conditions that a good entanglement measure $E[\hat{\rho}_{AB}]$ should satisfy are actually quite intuitive:

1. $E[\hat{\rho}_{AB}]$ is positive definite and equal to zero for separable states.
2. Given the mixture $\hat{\rho}_{AB} = \sum_j p_j \hat{\rho}_j$, where **p** is a probability distribution and $\{\hat{\rho}_j\}_j$ are density operators acting on $\mathcal{H}_A \otimes \mathcal{H}_B$, $E[\hat{\rho}_{AB}] \leqslant \sum_j p_j E[\hat{\rho}_j]$, which is to say that the entanglement of a collection of states cannot be increased by not knowing which one of them has been prepared, since this is corresponds to classical information.
3. At least on average, the entanglement level cannot increase when Alice and Bob apply protocols involving only LOCC.

These three conditions define what is known as an *entanglement monotone*. By themselves, they are not enough to define a unique entanglement measure even for pure states. However, by adding two more conditions known as *weak additivity* and *weak continuity* [3], which find an intuitive justification in the asymptotic limit of having infinitely many copies of the state, it is possible to prove that the *entanglement entropy* is the unique entanglement measure of pure bipartite states $|\psi\rangle_{AB}$. This measure is very intuitive, as it simply evaluates how mixed the reduced density operator of one of the parties remains after tracing out the other, that is, given the reduced density operators $\hat{\rho}_A = \text{tr}_B\{|\psi\rangle_{AB}\langle\psi|\}$ or $\hat{\rho}_B = \text{tr}_A\{|\psi\rangle_{AB}\langle\psi|\}$, this entanglement measure can be evaluated as

$$E[|\psi\rangle_{AB}] = S[\hat{\rho}_A] = S[\hat{\rho}_B]. \tag{2.5}$$

The equality of the von Neumann entropies of the reduced states will be clear after the following section. Hence, the problem of quantifying entanglement is basically solved for pure bipartite states. Pure states whose corresponding reduced states are maximally mixed are known as *maximally entangled states*.

The case of pure states allows for such a simple entanglement measure because all the correlations in the state are quantum, in the sense that they come from correlations at the interference level, as the only pure separable states are of the $|\psi_A\rangle \otimes |\psi_B\rangle$ form, which shows no interference terms between the partitions. The complication with mixed states, which can always be written from some ensemble decomposition $\{w_m, |\varphi_m\rangle_{AB}\}_{m=1,2,...,M}$ as explained in section 1.3.3, comes from the fact that part of the correlations can be due to the mixture of pure states according to the probability distribution **w**, correlations which will therefore have a classical character. Indeed, note that for mixed states the entanglement entropy is not even an entanglement monotone, as the entanglement entropy of the product state $\hat{\rho}_A \otimes \hat{\rho}_B$ is just the entropy of the mixed states $\hat{\rho}_A$ or $\hat{\rho}_B$, which is not zero except for pure states, and furthermore, depends on which mode is traced out. However, there are many quantifiers which are entanglement monotones, of which we discuss examples in the following:

- Possibly the most natural entanglement measure for mixed states is the *distillable entanglement* [8, 9]. Suppose that we give N copies of the mixed state $\hat{\rho}_{AB}$ to Alice and Bob. The process of *distillation* refers to the conversion of these copies to copies of maximally entangled states via protocols involving only LOCC. The distillable entanglement is defined as the maximum number of maximally entangled states that can be distilled from infinitely many copies of $\hat{\rho}_{AB}$. Apart from an entanglement monotone, it can also be shown to satisfy the weak additivity and weak continuity conditions, and to be equal to the entanglement entropy for pure states. Its drawback is that it requires a maximization over all the possible distillation protocols, and it is therefore very difficult to evaluate (paraphrasing [3]: it is a problem ranging from 'difficult' to 'hopeless'.)

- The *entanglement of formation* [8, 10] can be seen as the dual to the distillable entanglement: it measures the number of maximally entangled states that are needed to prepare infinitely many copies of the mixed state. It can be evaluated as

$$E_F[\hat{\rho}_{AB}] = \min_{\{w_k, |\psi_k\rangle\}_k} \sum_k w_k E[|\psi_k\rangle], \qquad (2.6)$$

where the minimization is performed over all the possible ensemble decompositions of $\hat{\rho}_{AB}$, what makes it a very hard measure to evaluate too. Nevertheless, closed formulas have been obtained for the entanglement of formation of the general state of two qubits ($d_A \times d_B = 2 \times 2$) [11], as well as for reduced classes of higher-dimensional bipartite states with strong symmetries [12].

- There is an entanglement monotone which can be evaluated fairly efficiently, as it does not require any optimization procedure: the *logarithmic negativity* [13, 14]. It quantifies how much the state $\hat{\rho}_{AB}$ violates the Peres–Horodecki criterion via

$$E_N[\hat{\rho}_{AB}] = \log\left\| \hat{\rho}_{AB}^{T_B} \right\|_1 = \log\left[1 + \sum_j \left(|\tilde{\lambda}_j| - \tilde{\lambda}_j \right) \right], \qquad (2.7)$$

where $\{\tilde{\lambda}_j\}_j$ are the eigenvalues of $\hat{\rho}_{AB}^{T_B}$, and $\|\hat{A}\|_1 = \mathrm{tr}\sqrt{\hat{A}\hat{A}^\dagger}$ denotes the so-called *trace norm*. The problem with this measure is that it does not collapse to the entanglement entropy (2.5) for pure states, and is not weakly additive in general.

2.3 Schmidt decomposition and purifications

When working with pure bipartite states, there is a very simple but powerful result known as the *Schmidt decomposition*, which says that this type of states can always be written in the form

$$|\psi\rangle = \sum_{j=1}^{d} \sqrt{\lambda_j} |u_j\rangle \otimes |v_j\rangle, \qquad (2.8)$$

where $d = \min\{d_A, d_B\}$, $\{\lambda_j\}_{j=1,2,\ldots,d}$ is a probability distribution (the $\sqrt{\lambda_j}$ are known as *Schmidt coefficients*), and $\{|u_j\rangle\}_{j=1,2,\ldots,d}$ and $\{|v_j\rangle\}_{j=1,2,\ldots,d}$ form orthonormal sets in $\mathcal{H}_A$ and $\mathcal{H}_B$, respectively. Note that this decomposition strikes right at the heart of entanglement as a 'superposition of joint states corresponding to mutually exclusive properties of the correlated systems'.

The proof of this result is very simple, so let us sketch it here. Consider two orthonormal bases $\{|a_m\rangle\}_{m=1,2,\ldots,d_A}$ and $\{|b_n\rangle\}_{n=1,2,\ldots,d_B}$ in $\mathcal{H}_A$ and $\mathcal{H}_B$, respectively. We can expand the state as

$$|\psi\rangle = \sum_{m=1}^{d_A} \sum_{n=1}^{d_B} C_{mn} |a_m\rangle \otimes |b_n\rangle. \qquad (2.9)$$

The expansion coefficients form a $d_A \times d_B$ matrix C, which according to the *singular-value decomposition theorem* [15] can always be written as $C = U\Lambda V^\dagger$. Here U and V are unitary[2] $d_A \times d_A$ and $d_B \times d_B$ matrices, respectively, and Λ is a diagonal $d_A \times d_B$ matrix with non-negative entries that we can then write as $\Lambda = \mathrm{diag}_{d_A \times d_B}(\sqrt{\lambda_1}, \sqrt{\lambda_2}, \ldots, \sqrt{\lambda_d})$. Introducing this decomposition in (2.9), or element-wise $C_{mn} = \sum_{j=1}^{d_A}\sum_{l=1}^{d_B}\sqrt{\lambda_j}\,U_{mj}\delta_{jl}(V^\dagger)_{ln} = \sum_{j=1}^{d}\sqrt{\lambda_j}\,U_{mj}V^*_{nj}$, we obtain

$$|\psi\rangle = \sum_{j=1}^{d}\sqrt{\lambda_j}\,\underbrace{\sum_{m=1}^{d}U_{mj}|a_m\rangle}_{|u_j\rangle} \otimes \underbrace{\sum_{n=1}^{d}V^*_{nj}|b_n\rangle}_{|v_j\rangle} \qquad (2.10)$$

which is exactly in the form (2.8). The orthonormality relations $\langle u_j|u_l\rangle = \delta_{jl} = \langle v_j|v_l\rangle$ follow directly from the unitarity of U and V, while the normalization of the *Schmidt distribution*, $\sum_{j=1}^{d}\lambda_j = 1$ follows from the normalization of the state, $\langle\psi|\psi\rangle = 1$.

With the state written in this form, it is completely trivial to evaluate the entanglement entropy: since the reduced density operators are diagonal, that is,

$$\hat{\rho}_A = \sum_{j=1}^{d}\lambda_j|u_j\rangle\langle u_j| \qquad \text{and} \qquad \hat{\rho}_B = \sum_{j=1}^{d}\lambda_j|v_j\rangle\langle v_j|, \qquad (2.11)$$

their von Neumann entropies give

$$E[|\psi\rangle] = -\sum_{j=1}^{d}\lambda_j \log \lambda_j. \qquad (2.12)$$

The number of non-zero Schmidt coefficients is known as the *Schmidt rank*. Separable states have then Schmidt rank equal to one. On the other hand, whenever $\lambda_j = 1/d \ \forall j$, the state will take the maximum entanglement possible $\log d$, corresponding to a maximally mixed reduced state in the Hilbert space with dimension d; under such conditions we then say that the system is in a *maximally entangled state*.

The Schmidt decomposition allows us to introduce the concept of *purification*: given a system A with associated Hilbert space $\mathcal{H}_A$ in a mixed state with diagonal representation $\hat{\rho}_A = \sum_{j=1}^{d}\lambda_j|r_j\rangle\langle r_j|$, we can always introduce another system B described by a Hilbert space $\mathcal{H}_B$ with the same dimension as $\mathcal{H}_A$, and with an orthonormal basis $\{|v_j\rangle\}_{j=1}^{d}$, and interpret the mixed state $\hat{\rho}_A$ as a reduction of the pure entangled state $|\psi\rangle_{AB} = \sum_{j=1}^{d}\sqrt{\lambda_j}|r_j\rangle \otimes |v_j\rangle$. Hence, a classical mixture of states can always be transformed into a pure entangled state in a 'doubled' Hilbert space.

---

[2] Note that in infinite dimensions C is a bounded matrix, since $|\psi\rangle$ must be normalized to one. Hence, a singular value decomposition still exists for it, but U and V are left-unitary.

Bibliography

[1] Einstein A, Podolsky B and Rosen N 1935 Can quantum-mechanical description of physical reality be considered complete? *Phys. Rev.* **47** 777

[2] Horodecki R, Horodecki P, Horodecki M and Horodecki K 2009 Quantum entanglement *Rev. Mod. Phys.* **81** 865

[3] Eisert J 2001 Entanglement in quantum information theory *PhD Thesis* arXiv: quant-ph/061025

[4] Eisert J and Plenio M B 2003 Introduction to the basics of entanglement theory in continuous-variable systems *Int. J. Quantum Inf.* **1** 479

[5] Nielsen M A and Chuang I L 2000 *Quantum Information and Quantum Computation* (Cambridge: Cambridge University Press)

[6] Peres A 1996 Separability criterion for density matrices *Phys. Rev. Lett.* **77** 1413

[7] Horodecki M, Horodecki P and Horodecki R 1996 Separability of mixed states: necessary and sufficient conditions *Phys. Lett.* A **223** 1

[8] Bennett C H, DiVincenzo D P, Smolin J A and Wootters W K 1996 Mixed-state entanglement and quantum error correction *Phys. Rev.* A **54** 3824

[9] Horodecki P and Horodecki R 2001 Distillation and bound entanglement *Quantum Inf. Comput.* **1** 45

[10] Wootters W K 2001 Entanglement of formation and concurrence *Quantum Inf. Comput.* **1** 27

[11] Wootters W K 1998 Entanglement of formation of an arbitrary state of two qubits *Phys. Rev. Lett.* **80** 2245

[12] Giedke G, Wolf M M, Krüger O, Werner R F and Cirac J I 2003 Entanglement of formation for symmetric Gaussian states *Phys. Rev. Lett.* **91** 107901

[13] Życzkowski K, Horodecki P, Sanpera A and Lewenstein M 1998 Volume of the set of separable states *Phys. Rev.* A **58** 883

[14] Vidal G and Werner R F 2002 Computable measure of entanglement *Phys. Rev.* A **65** 032314

[15] Bernstein D S 2005 *Matrix Mathematics: Theory Facts, and Formulas* (Princeton, NJ: Princeton University Press)

An Introduction to the Formalism of Quantum Information
with Continuous Variables

Carlos Navarrete-Benlloch

Chapter 3

Quantum operations

3.1 Basic principles of quantum operations

3.1.1 General considerations

Consider a system S with associated Hilbert space $\mathcal{H}_S$ subject to the actions of an experimentalist named Shane. *Quantum operations* refer, in essence[1], to the most general evolution that Shane can induce on the system S. As we explained in axioms 5 and 6, these surely involve unitary transformations and projective measurements[2] on the system, but are these the most general operations that Shane can apply?

The answer to this question is negative, since, as schematically shown in figure 3.1, Shane can always append an auxiliary system E with associated Hilbert space $\mathcal{H}_E$, apply unitaries and projective measurements on the joint system, and finally dismiss (trace out) the appended system E. Quantum operations then correspond to these joint unitaries and projective measurements as felt by the system S alone. In this context, it is customary to denote systems S and E by *system* and *environment*, respectively. We will refer to the combination of the system and the environment as the *joint system*.

As we will show in the last part of this section, any quantum operation can be represented by a map of the type [1, 2]

---

[1] In all fairness, they are not the *most* general type of evolution, since they do not capture some cases in which the system S interacts during the evolution with the systems with which it shared correlations initially, inducing a *non-completely positive map* on the state [1].

[2] We will refer to a measurement of a system observable as a *projective* measurement, to distinguish it from *generalized* measurements that we will study in the next section, which are not described by a set of projectors.

doi:10.1088/978-1-6817-4405-6ch3

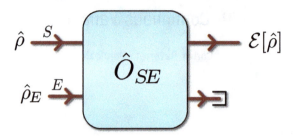

Figure 3.1. System–environment picture of quantum operations. The operator $\hat{O}_{SE}$ acting on the joint system can be a unitary or a projector, the latter being associated with a particular outcome in a measurement of a joint observable. The unitary or a non-selective measurement both lead to a trace-preserving quantum operation, while a selective measurement leads to a trace-decreasing one.

$$\mathcal{E}\left[\hat{Q}\right] = \sum_{k=1}^{K} \hat{E}_k \hat{Q} \hat{E}_k^{\dagger}, \tag{3.1}$$

where $\hat{Q}$ is any system operator, K can be selected at will by Shane, and the only restriction on the operators $\{\hat{E}_k\}_{k=1,2,\ldots,K}$ is

$$\sum_{k=1}^{K} \hat{E}_k^{\dagger} \hat{E}_k \leqslant \hat{I}, \tag{3.2}$$

where this type of operator inequalities, say $\hat{Q}_1 \leqslant \hat{Q}_2$, should be understood as $\hat{Q}_2 - \hat{Q}_1$ being positive semidefinite. If the state of the system was $\hat{\rho}$ prior to the quantum operation, it becomes

$$\hat{\rho}' = \frac{\mathcal{E}[\hat{\rho}]}{\mathrm{tr}\{\mathcal{E}[\hat{\rho}]\}}, \tag{3.3}$$

right after it. The operators $\{\hat{E}_k\}_{k=1,2,\ldots,K}$ are known as *Kraus operators*, and the expression (3.1) as an *operator-sum representation* of the quantum operation.

We can define two main types of quantum operations:

- Quantum operations which saturate (3.2) are known as *trace-preserving*, since $\mathrm{tr}\{\mathcal{E}[\hat{Q}]\} = \mathrm{tr}\{\hat{Q}\}$. As we will see in the next section, they correspond to operations that can be produced by performing a unitary transformation or a non-selective measurement on the joint system. Specifically, let us denote by $\hat{U}$ and $\hat{A} = \sum_{j=1}^{d_S \times d_E} a_j \hat{P}_j$ a unitary and an observable acting on the joint Hilbert space $\mathcal{H}_S \otimes \mathcal{H}_E$, respectively, $\{\hat{P}_j\}_{j=1,2,\ldots,d_S \times d_E}$ corresponding to the observable's eigenprojectors. Then the maps $\mathcal{E}_U[\hat{Q}] = \mathrm{tr}_E\{\hat{U}(\hat{Q} \otimes \hat{\rho}_E)\hat{U}^{\dagger}\}$ and $\mathcal{E}_A[\hat{Q}] = \mathrm{tr}_E\{\sum_{j=1}^{d_S \times d_E} \hat{P}_j(\hat{Q} \otimes \hat{\rho}_E)\hat{P}_j\}$, where $\hat{\rho}_E$ is some reference environmental state, correspond to trace-preserving quantum operations.

- Quantum operations which do not saturate (3.2) are known as *trace-decreasing*, since $\mathrm{tr}\{\mathcal{E}[\hat{Q}]\} < \mathrm{tr}\{\hat{Q}\}$. They require applying a selective measurement on the joint system, and the quantum operation is achieved only when a particular outcome appears (in other words, different outcomes correspond to different trace-decreasing operations). Explicitly, the maps $\mathcal{E}_j[\hat{Q}] = \text{tr}_E\{\hat{P}_j(\hat{Q} \otimes \hat{\rho}_E)\hat{P}_j\}$ correspond to trace-decreasing quantum operations, which occur with probability $\text{tr}\{\mathcal{E}_j[\hat{\rho}]\}$ if the system was in state $\hat{\rho}$ prior to the joint measurement.

Hence, trace-preserving quantum operations can be applied deterministically, while trace-decreasing ones only probabilistically. Trace-preserving quantum operations are also known as *quantum channels*, while the combination of all the trace-decreasing operations associated to a measurement on the joint system is known as a *generalized measurement*, of which we will tell more in the next section.

3.1.2 Further properties of quantum operations

The following important properties of quantum operations can be proved [1, 2]:
- Quantum operations can be defined *axiomatically*. In particular, it is possible to show that a map $\mathcal{E}[\hat{\rho}]$ acting on the space of density operators of the system admits an operator-sum representation of the type (3.1) if and only if
  1. It is a convex-linear map, that is, $\mathcal{E}[\sum_k w_k \hat{\rho}_k] = \sum_k w_k \mathcal{E}[\hat{\rho}_k]$, where $\{w_k\}_k$ is a probability distribution and $\{\hat{\rho}_k\}_k$ are density operators.
  2. $0 \leqslant \text{tr}\{\mathcal{E}[\hat{\rho}]\} \leqslant 1 \ \forall \hat{\rho}$.
  3. It is a completely positive (CP) map, that is, given a Hilbert space $\mathcal{H}_L$ of an arbitrary dimension, a product operator $\hat{O}_S \otimes \hat{O}_L$ and a positive-semidefinite operator $\hat{O}$ acting on the joint space $\mathcal{H}_S \otimes \mathcal{H}_L$, and the extended map defined by $\mathcal{E}_{\text{ext}}[\hat{O}_S \otimes \hat{O}_L] = \mathcal{E}[\hat{O}_S] \otimes \hat{O}_L$, then $\mathcal{E}_{\text{ext}}[\hat{O}]$ is also a positive-semidefinite operator for all $\hat{O}$.
- Two sets of Kraus operators $\{\hat{E}_k\}_{k=1,2,...,K}$ and $\{\hat{F}_m\}_{m=1,2,...,M}$ (we take $K \leqslant M$ for definiteness) lead to the same quantum operation if and only if there exists a left-unitary matrix U with elements $\{U_{mn}\}_{m,n=1,2,...,M}$ for which

$$\hat{E}_k = \sum_{m=1}^{M} U_{km}\hat{F}_m, \quad k = 1, 2,...,M, \tag{3.4}$$

where if $K \neq M$, then $M - K$ zeros must be included in the set with fewer Kraus operators, so that U is a square matrix.
- Any quantum operation has an operator-sum representation with $K \leqslant d_S^2$ Kraus operators.
- Any trace-preserving quantum operation can be written as the reduced unitary evolution of the joint system with an environment having $d_E \leqslant d_S^2$. Trace-decreasing operations require an extra measurement of a joint observable, and the desired quantum operation is accomplished only when a particular outcome appears in the measurement, which happens with a probability equal to the trace of the map (we say that it requires *post-selection*). The representation of the quantum operation in terms of unitaries and measurements acting onto the joint system is known as a *Stinespring dilation*.

3.1.3 Quantum operations as reduced dynamics in an extended system

In the rest of this section we will see explicitly how, after applying unitaries and measurements to the joint system, the reduced dynamics of the system is described by a map of the type (3.1). Let us start by writing the initial state of the joint system as

$$\hat{\rho}_{SE} = \hat{\rho} \otimes |\varphi_E\rangle\langle\varphi_E|. \tag{3.5}$$

Note that we do not lose generality by assuming the environmental state to be pure, since if it is mixed, we can always purify it by adding an extra environment, and taking both environments as a new joint environment.

If Shane applies a joint unitary $\hat{U}$ the state evolves into

$$\hat{\rho}_{SE}' = \hat{U}(\hat{\rho} \otimes |\varphi_E\rangle\langle\varphi_E|)\hat{U}^\dagger. \tag{3.6}$$

Introducing now an orthonormal basis in $\mathcal{H}_E$ given by $\{|e_k\rangle\}_{k=1,2,\ldots,d_E}$, the reduced state of the system is written as

$$\hat{\rho}' = \text{tr}_E\left\{\hat{\rho}_{SE}'\right\} = \sum_{k=1}^{d_E}\langle e_k|\hat{U}(\hat{\rho} \otimes |\varphi_E\rangle\langle\varphi_E|)\hat{U}^\dagger|e_k\rangle = \sum_{k=1}^{d_E}\hat{E}_k\hat{\rho}\hat{E}_k^\dagger, \tag{3.7}$$

where the operators $\hat{E}_k = \langle e_k|\hat{U}|\varphi_E\rangle$ act onto $\mathcal{H}_S$. Note that $\text{tr}_S\{\hat{\rho}'\} = 1 \;\forall\hat{\rho}$, and hence, the reduced unitary corresponds to a trace-preserving map such as (3.1) with a number of Kraus operators K given by the dimension of the Hilbert space of the environment.

If, on the other hand, Shane applies a non-selective measurement of a joint observable with diagonal representation

$$\hat{A} = \sum_{j=1}^{d_S \times d_E} a_j\hat{P}_j, \tag{3.8}$$

where $\{\hat{P}_j = |a_j\rangle\langle a_j|\}_{j=1,2,\ldots,d_S \times d_E}$ are the projectors onto its eigenvectors $\{|a_j\rangle \in \mathcal{H}_s \otimes \mathcal{H}_E\}_{j=1,2,\ldots,d_S \times d_E}$, the state evolves into

$$\hat{\rho}_{SE}' = \sum_{j=1}^{d_S \times d_E} \hat{P}_j\left(\hat{\rho} \otimes |\varphi_E\rangle\langle\varphi_E|\right)\hat{P}_j. \tag{3.9}$$

The reduced state of the system can then be written as

$$\hat{\rho}' = \text{tr}_E\left\{\hat{\rho}_{SE}'\right\} = \sum_{j=1}^{d_S \times d_E}\sum_{k=1}^{d_E}\langle e_k|\hat{P}_j\left(\hat{\rho} \otimes |\varphi_E\rangle\langle\varphi_E|\right)\hat{P}_j|e_k\rangle = \sum_{j=1}^{d_S \times d_E}\sum_{k=1}^{d_E}\hat{E}_{jk}\hat{\rho}\hat{E}_{jk}^\dagger, \tag{3.10}$$

where $\hat{E}_{jk} = \langle e_k|\hat{P}_j|\varphi_E\rangle$. Again, it is immediate to check that $\text{tr}_S\{\hat{\rho}'\} = 1 \;\forall\hat{\rho}$, and hence, the reduced non-selective measurement is a trace-preserving quantum operation with a number of Kraus operators given by $d_S \times d_E^2$.

The reduced dynamics of a selective measurement of the joint observable $\hat{A}$ is a little more subtle. Assume that after the measurement Shane obtains the outcome a_j, so that, accordingly, the state $\hat{\rho}_{SE}$ collapses to

$$\hat{\rho}_{SE,j} = p_j^{-1} \hat{P}_j \left(\hat{\rho} \otimes |\varphi_E\rangle\langle\varphi_E| \right) \hat{P}_j, \tag{3.11}$$

where

$$p_j = \mathrm{tr}\left\{ \hat{P}_j \hat{\rho}_{SE} \right\}, \tag{3.12}$$

is the probability for the outcome a_j to appear after the measurement. In this case it is easy to rewrite the reduced state of the system as

$$\hat{\rho}_j = \mathrm{tr}_E\left\{ \hat{\rho}_{SE,j} \right\} = p_j^{-1} \sum_{k=1}^{d_E} \hat{E}_{jk} \hat{\rho} \hat{E}_{jk}^\dagger, \tag{3.13}$$

and the corresponding probability as

$$p_j = \mathrm{tr}_S\left\{ \left[\sum_{k=1}^{d_E} \hat{E}_{jk}^\dagger \hat{E}_{jk} \right] \hat{\rho} \right\} \leqslant 1, \tag{3.14}$$

where $\hat{E}_{jk} = \langle e_k | \hat{P}_j | \varphi_E \rangle$. Hence, similarly to the previous cases, the reduced selective measurement is described by a map of the type (3.1) with d_E Kraus operators, but the map is trace decreasing in general.

3.2 Generalized measurements and positive operator-valued measures

As explained in the previous section, the most general type of measurement that one can perform on a system within this framework is described by a collection of trace-decreasing quantum operations $\{\mathcal{E}_j\}_{j=1,2,\ldots,J>1}$ each with an associated set of Kraus operators $\{\hat{E}_{jk}\}_{k=1,2,\ldots,K_j}$, which forms a complete set, that is,

$$\sum_{j=1}^{J} \mathrm{tr}\left\{ \mathcal{E}_j[\hat{\rho}] \right\} = 1 \ \forall \hat{\rho} \in \text{density operators}. \tag{3.15}$$

Generalized measurements for which $K_j = 1 \ \forall j$, that is, all its associated quantum operations are described by a single Kraus operator,

$$\mathcal{E}_j[\hat{\rho}] = \hat{E}_j \hat{\rho} \hat{E}_j^\dagger, \tag{3.16}$$

with

$$\hat{E}_j^\dagger \hat{E}_j \leqslant \hat{I} \ \forall j \qquad \text{and} \qquad \sum_{j=1}^{J} \hat{E}_j^\dagger \hat{E}_j = \hat{I}, \tag{3.17}$$

are very special, because it can be shown that its simplest Stinespring dilation does not require a joint unitary, just a joint projective measurement. The set

$\{\hat{\Pi}_j = \hat{E}_j^\dagger \hat{E}_j\}_{j=1,2,\ldots,J}$ is known as a *positive operator-valued measure* (POVM), while the operators $\{\hat{E}_j\}_{j=1,2,\ldots,J}$ are called the *measurement operators*.

These generalized measurements are the closest ones to projective measurements [2]. The POVM $\{\hat{\Pi}_j\}_{j=1,2,\ldots,J}$ plays the role of the spectral decomposition of the measured observable, from which we obtain the probability of observing the outcome 'j' as

$$p_j = \text{tr}\{\hat{\Pi}_j \hat{\rho}\}, \tag{3.18}$$

for a pre-measurement state $\hat{\rho}$. On the other hand, the measurement operators $\{\hat{E}_j\}_{j=1,2,\ldots,J}$ (uniquely defined from the POVM up to a left-unitary transformation) play the role of the projectors, with a post-measurement state given by

$$\hat{\rho}_j = p_j^{-1} \hat{E}_j \hat{\rho} \hat{E}_j^\dagger, \tag{3.19}$$

if the measurement is selective, or

$$\hat{\rho}' = \sum_{j=1}^{J} \hat{E}_j \hat{\rho} \hat{E}_j^\dagger, \tag{3.20}$$

if it is non-selective.

As a simple application of POVM-based measurements, consider the following problem. Suppose that someone picks one state out of the set $\{|\varphi_1\rangle, |\varphi_2\rangle\}$ and asks us to find with a single measurement which one was picked. If the states are orthogonal, this is trivial: we make a projective measurement defined by the projectors $\{\hat{P}_1 = |\varphi_1\rangle\langle\varphi_1|, \hat{P}_2 = |\varphi_2\rangle\langle\varphi_2|\}$, and check which outcome appeared. The problem is that it is simple to prove that when the states are not orthogonal, there is no strategy based on projective measurements allowing us to determine which state was given to us. However, we can design a strategy based on POVMs which will allow us to perform the needed task, although it does not work all the time.

Consider the POVM $\{\hat{\Pi}_1 = \hat{I} - |\varphi_2\rangle\langle\varphi_2|, \hat{\Pi}_2 = \hat{I} - |\varphi_1\rangle\langle\varphi_1|, \hat{\Pi}_3 = \hat{I} - \hat{\Pi}_1 - \hat{\Pi}_2\}$. Suppose that we obtain the outcome '1'; then, we know for sure that we obtained the state $|\varphi_1\rangle$, because the probability of observing '1' when the state is $|\varphi_2\rangle$ is zero, that is, $\langle\varphi_2|\hat{\Pi}_1|\varphi_2\rangle = 0$. The opposite happens when we get the outcome '2', we know for sure that $|\varphi_2\rangle$ was given to us, because $\langle\varphi_1|\hat{\Pi}_2|\varphi_1\rangle = 0$. Finally, when we get the outcome '3' we do not know which state we had, but at least we never make a misidentification of the state.

3.3 Local operations and classical communication protocols

We discuss in this section a very important class of operations performed on bipartite systems of the type discussed in the previous chapter. Suppose that Alice and Bob are in distant locations, so that one does not have access to the part of the system belonging to the other. In this scenario, it is natural to think that the most general class of operations that can be performed on the joint system are local operations (arbitrary quantum operations acting only on A or B independently) in a

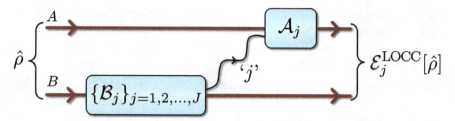

Figure 3.2. Prototypical form of a one-way-direct protocol involving only LOCC. Bob performs a generalized measurement and communicates the outcome to Alice, who applies a different trace-preserving operation depending on the information she receives from Bob.

correlated fashion (Alice and Bob can communicate by phone to decide together what to do). Quantum operations of this kind are known as *local operations and classical communication* (LOCC) *protocols*, and play a central role in many problems of quantum information (we already saw one, the characterization of entangled states).

As an example, consider the following prototypical protocol, schematically shown in figure 3.2: Bob performs a generalized measurement described by the set $\{\mathcal{B}_j\}_{j=1,2,\ldots,J}$ of trace-decreasing quantum operations, and communicates the outcome to Alice, who applies a trace-preserving quantum operation $\mathcal{A}_j$ from a pre-agreed set $\{\mathcal{A}_j\}_{j=1,2,\ldots,J}$ in one-to-one correspondence with the possible outcomes of Bob's measurements. Let us denote by $\{\hat{B}_{jk}\}_{k=1,2,\ldots,K_j}$ the Kraus operators associated with $\mathcal{B}_j$ (note that when $K_j = 1 \ \forall j$, Bob's measurement is a POVM-based measurement), and by $\{\hat{A}_{jm}\}_{m=1,2,\ldots,M_j}$ the ones associated with $\mathcal{A}_j$. The resulting possible maps will be given by

$$\mathcal{E}_j^{\mathrm{LOCC}}[\hat{\rho}] = \mathcal{A}_j[\mathcal{B}_j[\hat{\rho}]] = \sum_{k=1}^{K_j} \sum_{m=1}^{M_j} \left(\hat{A}_{jm} \otimes \hat{I}\right)\left(\hat{I} \otimes \hat{B}_{jk}\right)\hat{\rho}\left(\hat{I} \otimes \hat{B}_{jk}^{\dagger}\right)\left(\hat{A}_{jm}^{\dagger} \otimes \hat{I}\right). \quad (3.21)$$

We will find this type of LOCC protocols along the book (starting in the next section), which we will call *one-way-direct* LOCC protocols.

It is possible to prove that trace-preserving LOCC protocols can only decrease the entanglement of the state shared by Alice and Bob, as intuition says. What is a little more surprising is that trace-decreasing LOCC protocols can enhance the entanglement (we shall find one example of this when studying photon addition and subtraction), which is a further example of how counterintuitive quantum mechanics can be. Of course, on average a complete set of trace-decreasing LOCC maps (whose average can be seen as a trace-preserving LOCC protocol involving non-selective measurements) can only decrease the entanglement, showing that any local operation able to enhance the entanglement must be *intrinsically probabilistic*.

3.4 Majorization in quantum mechanics

In this section we introduce a relation between classical probability distributions called *majorization* [3, 4], which we will show to be connected to a couple of important questions in quantum mechanics [1], namely the freedom in the choice of ensemble decompositions of mixed states and the conversion of pure entangled states via LOCC protocols.

3.4.1 The concept of majorization

Majorization appeared as a way to order vectors or probability distributions in terms of their disorder, in an effort to understand when one can be built from another by randomizing the latter [3, 4].

Take two probability distributions $\mathbf{p} = \mathrm{col}(p_1, p_2,...,p_d)$ and $\mathbf{q} = \mathrm{col}(q_1, q_2,...,q_d)$, where d can be infinite (see an example with $d = 13$ in figure 3.3). We say that $\mathbf{p}$ *majorizes* $\mathbf{q}$, and denote it by $\mathbf{p} \succ \mathbf{q}$, if and only if

$$\sum_{n=1}^{m} p_n^{\downarrow} \geqslant \sum_{n=1}^{m} q_n^{\downarrow}, \forall\, m < d, \tag{3.22}$$

where $\mathbf{p}^{\downarrow}$ and $\mathbf{q}^{\downarrow}$ are the original vectors with their components rearranged in decreasing order.

This characterization of the majorization relation is interesting from an operational point of view, since it is easy to check numerically whether two vectors satisfy this condition or not. Nevertheless, it can be proven that $\mathbf{p} \succ \mathbf{q}$ is strictly equivalent to two other relations:

- For every concave function $h(x)$, it is satisfied $\sum_{n=1}^{d} h(p_n) \leqslant \sum_{n=1}^{d} h(q_n)$.
- $\mathbf{q}$ can be obtained from $\mathbf{p}$ as $\mathbf{q} = D\mathbf{p}$, where D is a column-stochastic matrix[3].

These relations are very interesting from an interpretational point of view. First, note that the Shannon entropy is a concave function, and hence, the first relation says that the entropy of $\mathbf{q}$ is larger than the entropy of $\mathbf{p}$; now, as we have discussed in previous chapters, larger entropy means larger lack of information, which can be somehow interpreted as more disorder. Second, it is a known result that any column-stochastic matrix can be written as a convex sum (or statistical mixture) of permutations; hence, the second relation says that $\mathbf{q}$ can be obtained from $\mathbf{p}$ by applying a random mixture of permutations on the latter. Hence, both conditions seem to justify the idea of $\mathbf{q}$ being more disordered than $\mathbf{p}$.

3.4.2 Majorization and ensemble decompositions of a state

As a first simple application of majorization to quantum mechanics, we answer the following question: under which conditions can we find an ensemble decomposition based on a probability distribution $\mathbf{w}$, say $\{w_m, |\varphi_m\rangle\}_{m=1,2,...,M}$, of a density operator $\hat{\rho}$ with diagonal representation $\hat{\rho} = \sum_{n=1}^{d} \lambda_n |r_n\rangle\langle r_n|$?

---

[3] A square matrix is *column-stochastic* when its elements are real and positive, each of its columns sum to one, and each of its rows sum to less than one. Most of the literature on the connection between majorization and quantum information studies finite-dimensional systems, in which case it can be shown that column-stochastic matrices are also *doubly stochastic* (all columns and rows sum to one). One needs the slightly more general definition of column-stochastic to cope with infinite-dimensional spaces [4], as we will do in the next chapter.

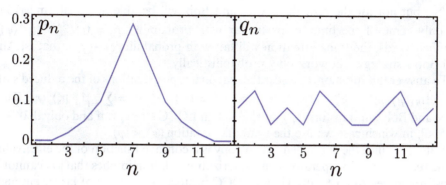

Figure 3.3. Examples of probability distributions in $d = 13$ dimensions. Clearly **p** is more ordered than **q**. What is not immediate to see is whether **q** can be obtained by randomizing **p**, and here is where majorization theory helps.

The answer is rather simple: only when $\lambda \succ \mathbf{w}$ [1]. Of course, if $M \neq d$, zeros are added in the vector with smaller dimensionality to match the dimensions of λ and $\mathbf{w}$.

This result is indeed a simple consequence of what we learned in section 1.3.3 concerning the freedom to represent a given density operator by different ensemble decompositions. In particular, if $\hat{\rho}$ can be represented by the ensembles $\{w_m, |\varphi_m\rangle\}_{m=1,2,...,M}$ and $\{\lambda_n, |r_n\rangle\}_{n=1,2,...,d}$, there must exist a left-unitary matrix U with elements $\{U_{mn}\}_{m,n=1,2,...,\max\{M,d\}}$ for which

$$\sqrt{w_m}|\varphi_m\rangle = \sum_{n=1}^{\max\{M,d\}} U_{mn}\sqrt{\lambda_n}|r_n\rangle, \quad m = 1, 2,..., \max\{M, d\}. \quad (3.23)$$

Now, taking the inner product of this expression with itself, and using the orthonormality of the $\{|r_n\rangle\}_n$ set, we obtain

$$w_m = \sum_{n=1}^{\max\{M,d\}} |U_{mn}|^2 \lambda_n, \quad m = 1, 2,..., \max\{M, d\}. \quad (3.24)$$

Finally, as U is left-unitary, the matrix with elements $\{|U_{mn}|^2\}_{m,n=1,2,...,\max\{M,d\}}$ is column-stochastic.

3.4.3 Majorization and the transformation of entangled states

The next application of majorization theory to quantum mechanics appears when answering another question of paramount importance in information theory: given a bipartite state $|\psi\rangle \in \mathcal{H}_A \otimes \mathcal{H}_B$, under which conditions can it be transformed *deterministically* into another state $|\varphi\rangle \in \mathcal{H}_A \otimes \mathcal{H}_B$ if Alice and Bob are allowed to use only LOCC protocols?

By 'deterministically' we mean that given a complete set of LOCC protocols $\{\mathcal{E}_j^{\text{LOCC}}\}_j$, the transformation $|\psi\rangle \to |\varphi\rangle$ succeeds for all of them, that is, $\mathcal{E}_j^{\text{LOCC}}[|\psi\rangle\langle\psi|] \propto |\varphi\rangle\langle\varphi| \; \forall j$. If, for example, the transformation works only for

$\mathcal{E}_0^{\mathrm{LOCC}}$, but not for the rest, then Alice and Bob will be able to transform $|\psi\rangle$ into $|\varphi\rangle$ only some of the time, in particular with probability $p_0 = \mathrm{tr}\{\mathcal{E}_0^{\mathrm{LOCC}}[|\psi\rangle\langle\psi|]\}$. In other words, the transformation will fail with probability $(1 - p_0)$, that is, Alice and Bob's strategy will work only probabilistically.

To answer this question, consider the diagonal representations of the reduced states $\hat{\rho}_A^{\psi} = \mathrm{tr}_B\{|\psi\rangle\langle\psi|\} = \sum_{n=1}^{d}\lambda_n^{\psi}\,|r_n\rangle_{\psi}\langle r_n|$ and $\hat{\rho}_A^{\varphi} = \mathrm{tr}_B\{|\varphi\rangle\langle\varphi|\} = \sum_{n=1}^{d}\lambda_n^{\varphi}\,|r_n\rangle_{\varphi}\langle r_n|$. Then, Alice and Bob can transform $|\psi\rangle$ into $|\varphi\rangle$ via an LOCC strategy, if and only if $\lambda^{\psi} \prec \lambda^{\varphi}$ [1, 5, 6], in which case we use the symbolic notation $|\psi\rangle \prec |\varphi\rangle$.

Note that since the entanglement entropy is a concave function of the eigenvalues of the reduced density operator, this majorization relation implies that $|\psi\rangle$ cannot be transformed deterministically via an LOCC protocol into states of larger entanglement, that is, $E[|\psi\rangle] \geqslant E[|\varphi\rangle]$, as expected.

It is also possible to prove that if $|\psi\rangle$ can be transformed into $|\varphi\rangle$ deterministically via an LOCC protocol, it can always be achieved with a one-way-direct LOCC protocol of the following simple form [6]: Bob performs a measurement described by some POVM $\{\hat{\Pi}_j\}_{j=1,2,\ldots,J}$ and communicates the outcome to Alice, who applies a unitary transformation from a pre-agreed set of unitaries $\{\hat{A}_j\}_{j=1,2,\ldots,J}$ in one to one correspondence with the possible outcomes of Bob's measurement. Hence, if $\{\hat{B}_j\}_{j=1,2,\ldots,J}$ are the measurement operators associated to Bob's POVM, the transformation is accomplished as

$$|\varphi\rangle \propto \left(\hat{A}_j \otimes \hat{I}\right)\left(\hat{I} \otimes \hat{B}_j\right)|\psi\rangle \; \forall j. \tag{3.25}$$

Bibliography

[1] Nielsen M A and Chuang I L 2000 *Quantum Information and Quantum Computation* (Cambridge: Cambridge University Press)

[2] Paris M G A 2012 The modern tools of quantum mechanics: a tutorial on quantum states, measurements, and operations *Eur. Phys. J. Spec. Top.* **203** 61

[3] Arnold B 1987 Lecture Notes in Statistics *Majorization and the Lorenz Order* vol 43 (Berlin: Springer)

[4] Kaftal V and Weiss G 2010 An infinite dimensional Schur-Horn theorem and majorization theory *J. Funct. Anal.* **259** 3115

[5] Nielsen M A 1999 Conditions for a class of entanglement transformations *Phys. Rev. Lett.* **83** 436

[6] Nielsen M A and Vidal G 2001 Majorization and the interconversion of bipartite states *Quantum Inf. Comput.* **1** 76

An Introduction to the Formalism of Quantum Information with Continuous Variables

Carlos Navarrete-Benlloch

Chapter 4

Quantum information with continuous variables

4.1 The classical harmonic oscillator

Consider the basic mechanical model of a *one-dimensional harmonic oscillator*: a particle of mass m is at rest at some equilibrium position which we take as $x = 0$; when displaced from this position by some amount a, a restoring force $F = -kx$ starts acting on the particle, trying to bring it back to $x = 0$. Newton's equation of motion for the particle is therefore $m\ddot{x} = -kx$, which together with the initial conditions $x(0) = a$ and $\dot{x}(0) = v$ gives the solution $x(t) = a \cos \omega t + (v/\omega)\sin \omega t$, $\omega = \sqrt{k/m}$ being the so-called *angular frequency*. Therefore the particle will be bouncing back and forth between positions $-\sqrt{a^2 + v^2/\omega^2}$ and $\sqrt{a^2 + v^2/\omega^2}$ with time period $2\pi/\omega$ (hence the name 'harmonic oscillator').

Let us now study the problem from a Hamiltonian point of view. For this one-dimensional problem with no constraints, we can take the position of the particle and its momentum as the generalized coordinate and momentum, that is, $q = x$ and $p = m\dot{x}$. The restoring force derives from a potential $V(x) = kx^2/2$, and hence the Hamiltonian takes the form

$$H = \frac{p^2}{2m} + \frac{m\omega^2}{2}q^2. \tag{4.1}$$

The canonical equations read

$$\dot{q} = \frac{p}{m} \qquad \text{and} \qquad \dot{p} = -m\omega^2 q, \tag{4.2}$$

which together with the initial conditions $q(0) = a$ and $p(0) = mv$ give the trajectory

$$\left(q, \frac{p}{m\omega}\right) = \left(a \cos \omega t + \frac{v}{\omega} \sin \omega t, \frac{v}{\omega} \cos \omega t - a \sin \omega t\right), \tag{4.3}$$

doi:10.1088/978-1-6817-4405-6ch4

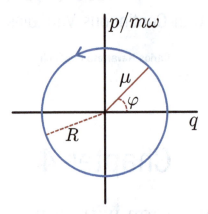

Figure 4.1. Phase-space trajectory of the classical harmonic oscillator. Both the real and complex phase-space variables, (x, p) and (μ, φ), respectively, are shown.

where we normalize the momentum by $m\omega$ for simplicity. Starting at the phase-space point $(a, v/\omega)$ the system evolves periodically drawing a circle of radius $R = \sqrt{a^2 + v^2/\omega^2}$ as shown in figure 4.1, returning to its initial point at times $t_k = 2\pi k/\omega$, with $k \in \mathbb{N}$. This circular trajectory could have been derived without even solving the equations of motion, as the conservation of the Hamiltonian $H(t) = H(0)$ leads directly to $q^2 + p^2/m^2\omega^2 = R^2$, which is exactly the circumference of figure 4.1. This is a simple manifestation of the power of the Hamiltonian formalism.

There is another useful description of the harmonic oscillator, the so-called *amplitude–phase* or *complex* representation. The amplitude and phase refer to the polar coordinates in phase space, say $\mu = \sqrt{q^2 + p^2/m^2\omega^2}$ and $\varphi = \arctan(p/m\omega q)$, as shown in figure 4.1. In terms of these variables, the trajectory reads simply $(\mu, \varphi) = (R, \varphi_0 - \omega t)$, with $\varphi_0 = \arctan(v/\omega a)$, so that the evolution is completely described by a linear time variation of the oscillator's phase. From these variables we can define the complex variable $v = \mu \exp(i\varphi) = q + (i/m\omega)p$, in terms of which the trajectory reads $v(t) = R \exp[i(\varphi_0 - \omega t)]$, and the Hamiltonian can be written as $H = m\omega^2 v^* v/2$. The complex variable v is known as the *normal variable* of the oscillator.

4.2 The quantum harmonic oscillator

The *harmonic oscillator* is the prototype of a system described quantum mechanically by an infinite-dimensional Hilbert space. In order to see this, let us find the eigenstates of its Hamiltonian, which is given by the operator

$$\hat{H} = \frac{\hat{p}^2}{2m} + \frac{m\omega^2}{2}\hat{q}^2, \tag{4.4}$$

by virtue of the discussion after axiom 3, with the *position* $\hat{q}$ and *momentum* $\hat{p}$ satisfying the commutation relation

$$[\hat{q}, \hat{p}] = i\hbar. \tag{4.5}$$

We will always work with dimensionless versions of them, the so-called X and P *quadratures* (although we may keep using the names 'position' and 'momentum' most of the time)

$$\hat{X} = \sqrt{\frac{2\omega m}{\hbar}}\,\hat{q} \qquad \text{and} \qquad \hat{P} = \sqrt{\frac{2}{\hbar\omega m}}\,\hat{p}, \tag{4.6}$$

which satisfy the commutation relation

$$\left[\hat{X}, \hat{P}\right] = 2\mathrm{i}, \tag{4.7}$$

and therefore the uncertainty relations

$$\Delta X \Delta P \geqslant 1. \tag{4.8}$$

In terms of these quadratures, the Hamiltonian reads

$$\hat{H} = \frac{\hbar\omega}{4}\left(\hat{X}^2 + \hat{P}^2\right). \tag{4.9}$$

In order to find the eigensystem of this operator, we decompose the quadratures as

$$\hat{X} = \hat{a}^\dagger + \hat{a} \qquad \text{and} \qquad \hat{P} = \mathrm{i}(\hat{a}^\dagger - \hat{a}), \tag{4.10}$$

where the operators $\hat{a}$ and $\hat{a}^\dagger$, known as the *annihilation* and *creation operators*, satisfy the commutation relation

$$[\hat{a}, \hat{a}^\dagger] = 1. \tag{4.11}$$

In terms of these operators, the Hamiltonian is rewritten as

$$\hat{H} = \hbar\omega(\hat{a}^\dagger\hat{a} + 1/2), \tag{4.12}$$

and hence the problem has been reduced to finding the eigensystem of the so-called *number operator* $\hat{N} = \hat{a}^\dagger\hat{a}$.

Let us denote by n a generic real number contained in the spectrum of $\hat{N}$, whose corresponding eigenvector we denote by $|n\rangle$, so that, $\hat{N}|n\rangle = n|n\rangle$. The eigensystem of $\hat{N}$ is readily found from the following two properties:

- $\hat{N}$ is a positive-semidefinite operator, as for any vector $|\psi\rangle$ it is satisfied $\langle\psi|\hat{N}|\psi\rangle = (\hat{a}|\psi\rangle, \hat{a}|\psi\rangle) \geqslant 0$. When applied to its eigenvectors, this property forbids the existence of negative eigenvalues, that is, $n \geqslant 0$.
- Applying the commutation relation[1] $[\hat{N}, \hat{a}] = -\hat{a}$ to $|n\rangle$, it is straightforward to show that the vector $\hat{a}|n\rangle$ is also an eigenvector of $\hat{N}$ with eigenvalue $n - 1$. Similarly, from the commutation relation $[\hat{N}, \hat{a}^\dagger] = \hat{a}^\dagger$ it is found that the vector $\hat{a}^\dagger|n\rangle$ is an eigenvector of $\hat{N}$ with eigenvalue $n + 1$.

---

[1] This is straightforward to find by using the property $[\hat{A}\hat{B}, \hat{C}] = \hat{A}[\hat{B}, \hat{C}] + [\hat{A}, \hat{C}]\hat{B}$, valid for any three operators $\hat{A}$, $\hat{B}$, and $\hat{C}$.

These two properties imply that the spectrum of $\hat{N}$ is the set of non-negative integers $n = 0, 1, 2, \ldots$, and that the eigenvector $|0\rangle$ corresponding to $n = 0$ must satisfy $\hat{a}|0\rangle = 0$. Otherwise, it would be possible to find negative eigenvalues, hence contradicting the positivity of $\hat{N}$. Thus, the set of eigenvectors $\{|n\rangle\}_{n=0,1,\ldots}$ is an infinite, countable set. Moreover, note that $\langle n|\hat{N}|m\rangle = n\langle n|m\rangle = m\langle n|m\rangle$ implies that the eigenvectors form an orthogonal set, that is, $\langle n|m\rangle = 0$ for $n \neq m$. In addition, we will show later by using a specific representation, see (4.27), that the eigenstate $|0\rangle$ can be normalized to one, which implies that the rest of the eigenstates can be normalized as well by virtue of (4.13). Therefore, the eigenstates of $\hat{N}$ form an orthonormal set, that is, $\langle n|m\rangle = \delta_{nm}$. Finally, according to the axioms of quantum mechanics only the vectors normalized to one are physically relevant for the description of the state of the harmonic oscillator, and hence we conclude that the vector space spanned by the eigenvectors of $\hat{N}$ is an infinite-dimensional Hilbert space, since it is isomorphic to $l^2(\infty)$, see section 1.2.3.

Summarizing, we have been able to prove that the Hilbert space associated with the one-dimensional harmonic oscillator is infinite-dimensional. In the process, we have explicitly built an orthonormal basis of this space by using the eigenvectors $\{|n\rangle\}_{n=0,1,\ldots}$ of the number operator $\hat{N}$, known as the *Fock basis*. The annihilation and creation operators, $\hat{a}$ and $\hat{a}^\dagger$, allow us to move through this basis as

$$\hat{a}|n\rangle = \sqrt{n}|n-1\rangle \qquad \text{and} \qquad \hat{a}^\dagger|n\rangle = \sqrt{n+1}|n+1\rangle, \qquad (4.13)$$

the factors in the square roots being easily found from normalization requirements. The state $|0\rangle$ is known as the *vacuum* state, since it corresponds to the eigenstate of $\hat{N}$ with no *excitations*, $n = 0$.

In contrast to the number operator, which has a discrete spectrum, the quadrature operators possess a pure continuous spectrum. Let us focus on the $\hat{X}$ operator, whose eigenvectors we denote by $\{|x\rangle\}_{x\in\mathbb{R}}$ with corresponding eigenvalues $\{x\}_{x\in\mathbb{R}}$, that is,

$$\hat{X}|x\rangle = x|x\rangle. \qquad (4.14)$$

In order to prove that $\hat{X}$ has a pure continuous spectrum, just note that, from the relation

$$\exp\left(\frac{i}{2}y\hat{P}\right)\hat{X}\exp\left(-\frac{i}{2}y\hat{P}\right) = \hat{X} + y, \qquad (4.15)$$

which is easily found via the Baker–Campbell–Haussdorf lemma[2], it follows that if $|x\rangle$ is an eigenvector of $\hat{X}$ with x eigenvalue, then the vector $\exp(-iy\hat{P}/2)|x\rangle$ is also an eigenvector of $\hat{X}$ with eigenvalue $x + y$. Now, as this holds for any real y, we

---

[2] This lemma reads

$$e^{\hat{B}}\hat{A}e^{-\hat{B}} = \sum_{n=0}^{\infty}\frac{1}{n!}\underbrace{\left[\hat{B},\left[\hat{B},\ldots\left[\hat{B},\hat{A}\right]\ldots\right]\right]}_{n}, \qquad (4.16)$$

and is valid for two general operators $\hat{A}$ and $\hat{B}$.

conclude that the spectrum of $\hat{X}$ is the whole real line. Moreover, as a self-adjoint operator, one can use its eigenvectors as a continuous basis of the Hilbert space of the oscillator by using the Dirac normalization $\langle x|y \rangle = \delta(x - y)$. The same results can be obtained for the $\hat{P}$ operator, whose eigenvectors we denote by $\{|p\rangle\}_{p \in \mathbb{R}}$ with corresponding eigenvalues $\{p\}_{p \in \mathbb{R}}$, that is,

$$\hat{P}|p\rangle = p|p\rangle. \tag{4.17}$$

Note that these results rely only on the canonical commutation relations, and hence are completely general, valid for any system, not only for the harmonic oscillator. Note also that not being vectors contained in the Hilbert space of the oscillator (they cannot be properly normalized), the position and momentum eigenvectors cannot correspond to physical states. Nevertheless, we will see that they can be understood as an unphysical limit of some physical states (the squeezed states).

It is not difficult to prove that there exists a Fourier transform relation between the position and momentum bases, that is,

$$|p\rangle = \int_{-\infty}^{+\infty} \frac{\mathrm{d}x}{\sqrt{4\pi}} \exp\left(\frac{i}{2}px\right) |x\rangle \iff |x\rangle = \int_{-\infty}^{+\infty} \frac{\mathrm{d}p}{\sqrt{4\pi}} \exp\left(-\frac{i}{2}px\right) |p\rangle. \tag{4.18}$$

To this aim we now prove that

$$\langle x|p \rangle = \frac{1}{\sqrt{4\pi}} \exp(ixp/2). \tag{4.19}$$

First note that the commutator $[\hat{X}, \hat{P}] = 2i$ implies that

$$\langle x|\hat{P}|x'\rangle = \frac{2i\delta(x - x')}{x - x'}, \tag{4.20}$$

and hence

$$\langle x|\hat{P}|\psi \rangle = \int_{\mathbb{R}} \mathrm{d}x' \langle x|\hat{P}|x'\rangle \langle x'|\psi\rangle = \int_{\mathbb{R}} \mathrm{d}x' \frac{2i\delta(x - x')}{x - x'} \langle x'|\psi\rangle$$

$$= \int_{\mathbb{R}} \mathrm{d}x' \frac{2i\delta(x - x')}{x - x'} \left[\langle x|\psi\rangle + (x' - x)\frac{\mathrm{d}\langle x|\psi\rangle}{\mathrm{d}x} + \sum_{n=2}^{\infty} \frac{(x' - x)^n}{n!} \frac{\mathrm{d}^n\langle x|\psi\rangle}{\mathrm{d}x^n} \right]. \tag{4.21}$$

The order zero of the Taylor expansion is zero because the kernel is antisymmetric around x, while the terms of order two or above give zero as well after integrating them. This means that

$$\langle x|\hat{P}|\psi \rangle = -2i\frac{\mathrm{d}\langle x|\psi\rangle}{\mathrm{d}x}, \tag{4.22}$$

which applied to $|\psi\rangle = |p\rangle$ yields the differential equation

$$p\langle x|p \rangle = -2i\frac{\mathrm{d}\langle x|p\rangle}{\mathrm{d}x}, \tag{4.23}$$

which has (4.19) as its solution, the factor $1/\sqrt{4\pi}$ coming from the Dirac normalization of the $|p\rangle$ vectors.

As an example of the use of these continuous representations, we now find the position representation of the number states, which we write as

$$|n\rangle = \int_{\mathbb{R}} \mathrm{d}x \psi_n(x) |x\rangle. \tag{4.24}$$

As a first step we find the projection of vacuum onto a position eigenstate, the so-called *ground state wave function* $\psi_0(x) = \langle x|0\rangle$, from

$$0 = \langle x|\hat{a}|0\rangle = \frac{1}{2}\langle x|(\hat{X} + i\hat{P})|0\rangle = \frac{1}{2}\left(x + 2\frac{\mathrm{d}}{\mathrm{d}x}\right)\psi_0(x), \tag{4.25}$$

where we have used (4.22), which is a differential equation for $\psi_0(x)$ having

$$\psi_0(x) = \frac{1}{(2\pi)^{1/4}} \exp\left(-\frac{x^2}{4}\right), \tag{4.26}$$

as its solution. The factor $(2\pi)^{-1/4}$ is found by imposing the normalization

$$\langle 0|0\rangle = \int_{\mathbb{R}} \mathrm{d}x \psi_0^2(x) = 1. \tag{4.27}$$

Now, the projection of any number state $|n\rangle$ onto a position eigenstate (the nth *excited wave function*) is found from the ground state wave function as

$$\psi_n(x) = \langle x|n\rangle = \frac{1}{\sqrt{n!}}\langle x|\hat{a}^{\dagger n}|0\rangle = \frac{1}{\sqrt{n!}\,2^n}\langle x|(\hat{X} - i\hat{P})^n|0\rangle = \frac{1}{\sqrt{n!}\,2^n}\left(x - 2\frac{\mathrm{d}}{\mathrm{d}x}\right)^n \psi_0(x), \tag{4.28}$$

which, recalling the Rodrigues formula for the Hermite polynomials

$$H_n\left(\frac{x}{\sqrt{2}}\right) = 2^{-n/2} \exp\left(\frac{x^2}{4}\right)\left(x - 2\frac{\mathrm{d}}{\mathrm{d}x}\right)^n \exp\left(-\frac{x^2}{4}\right), \tag{4.29}$$

leads to the simple expression

$$\psi_n(x) = \frac{1}{\sqrt{2^{n+1/2}\pi^{1/2}n!}} H_n\left(\frac{x}{\sqrt{2}}\right) \exp\left(-\frac{x^2}{4}\right). \tag{4.30}$$

Finally, let us stress that even though all that we are going to discuss in the following applies to a general bosonic system, that is, a system described by a collection of harmonic oscillators, we will always have in mind the electromagnetic field (light, in particular), which can be described as a set of *modes* with well defined polarization and spatio-temporal profile, each of which behaves as the mechanical harmonic oscillator that we have introduced. Consequently, we will use 'modes' and 'harmonic oscillators', as well as 'photons' and 'excitations' interchangeably. Nevertheless, it is important to keep in mind that there are many other physical systems which behave as quantum harmonic oscillators, for example, motional degrees of freedom in atoms or mesoscopic objects, polarized atomic ensembles, superconducting circuits, or even the degrees of freedom of fields other than the electromagnetic one, e.g. the ones associated to the weak and strong interactions or the Higgs boson.

4.3 The harmonic oscillator in phase space: the Wigner function

4.3.1 General considerations

As the position and momentum do not have common eigenstates and, moreover, their eigenvectors cannot correspond to physical states of the oscillator, one concludes that these observables cannot take definite values in quantum mechanics. Given the state $\hat{\rho}$, the best one can offer is the *probability density functions* which will dictate the statistics of a measurement of these observables, $P(x) = \langle x|\hat{\rho}|x\rangle$ and $P(p) = \langle p|\hat{\rho}|p\rangle$. In other words, quantum mechanically there are not well defined trajectories in phase space, the position and momentum of the oscillator are always affected by some (*quantum*) *noise*.

The following question arises naturally: is it then possible to describe quantum mechanics as a probability distribution in phase space which blurs the classical trajectories? As we are about to see, the answer is only partially positive, as quantum noise is much more subtle than common classical noise.

A logical way of building such a phase-space distribution, say $W_{\hat{\rho}}(x, p)$, is as the one having the probability density functions $P(x)$ and $P(p)$ as its marginals, that is,

$$P(x) = \int_{\mathbb{R}} \mathrm{d}p\, W_{\hat{\rho}}(x, p) \qquad \text{and} \qquad P(p) = \int_{\mathbb{R}} \mathrm{d}x\, W_{\hat{\rho}}(x, p). \qquad (4.31)$$

It is possible to show that this distribution is uniquely defined by [1]

$$W_{\hat{\rho}}(x, p) = \frac{1}{4\pi} \int_{\mathbb{R}} \mathrm{d}y\, \exp\left(-\frac{\mathrm{i}}{2}py\right)\langle x + y/2|\hat{\rho}|x - y/2\rangle, \qquad (4.32)$$

which is known as the *Wigner function*. It is immediate to check that this distribution has the proper marginals, and that it is normalized, i.e. $\int_{\mathbb{R}^2} \mathrm{d}x\mathrm{d}p\, W_{\hat{\rho}}(x, p) = 1$. The proof of its uniqueness is not that simple, however, see [1].

As we will prove in the next section, given the state $\hat{\rho}$ of the oscillator, the quantum expectation value of any operator can be found from

$$\left\langle \left(\hat{X}^m \hat{P}^n\right)^{(s)} \right\rangle = \int_{\mathbb{R}^2} \mathrm{d}x\mathrm{d}p\, W_{\hat{\rho}}(x, p)x^m p^n, \qquad (4.33)$$

where $(\hat{X}^m \hat{P}^n)^{(s)}$ refers to the symmetrized version of the corresponding moment with respect to position and momentum, e.g. $(\hat{X}^2 \hat{P})^{(s)} = (\hat{X}^2 \hat{P} + \hat{P}\hat{X}^2 + \hat{X}\hat{P}\hat{X})/3$.

Taking into account that the prescription to find the quantum operator associated with a classical observable $O(x, p)$ consists specifically of symmetrizing it with respect to x and p, and then changing the position and momentum by the corresponding self-adjoint operators (which guarantees the self-adjointness of the remaining operator, as we saw in section 1.3.5), this result reinforces the interpretation of $W_{\hat{\rho}}(x, p)$ as a probability distribution in phase space, and hence of quantum mechanics as noise acting on the classical trajectories. However, it is readily seen that this distribution can take on negative values for many quantum-mechanical states, and hence it is not a true two-dimensional probability density function. As an example of this, let us consider the Wigner function of the number states $\hat{\rho} = |n\rangle\langle n|$. Although not straightforward [1], it is possible to show that the corresponding Wigner function is given by

$$W_{|n\rangle}(x, p) = \frac{(-1)^n}{2\pi} L_n(x^2 + p^2) \exp\left(-\frac{x^2 + p^2}{2}\right), \tag{4.34}$$

where $L_n(z)$ is the Laguerre polynomial of order n, which can be found from the Rodrigues formula

$$L_n(z) = \frac{\exp(z)}{n!} \frac{\mathrm{d}^n}{\mathrm{d}z^n}[z^n \exp(-z)]. \tag{4.35}$$

For any $n > 0$, this function has negative regions and, therefore, it cannot be simulated with any source of classical noise. For example, for odd n it is always negative at the origin of phase space, since $L_n(0) = 1 \; \forall \, n$. The Wigner functions of the first four Fock states are plotted in figure 4.2.

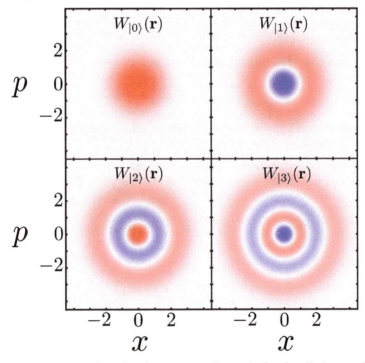

Figure 4.2. Density plot of the Wigner functions corresponding to the first four Fock states. Red and blue regions correspond to positive and negative values of the function, respectively. In both cases regions with higher contrast correspond to larger absolute values.

4.3.2 Definition based on the characteristic function

Given a state $\hat{\rho}$ of the oscillator, expression (4.32) allows us to compute the corresponding Wigner function. However, there is a much more convenient way of writing the Wigner distribution, which is based on the so-called *displacement operator*

$$\hat{D}(\mathbf{r}) = \exp\left[\frac{i}{2}\hat{\mathbf{R}}^T \Omega \mathbf{r}\right], \tag{4.36}$$

where $\mathbf{r} = \text{col}(x, p)$ is the coordinate vector in phase space (which is known as the *displacement* induced by the operator), $\hat{\mathbf{R}} = \text{col}(\hat{X}, \hat{P})$ is the corresponding vector operator, and $\Omega = \begin{bmatrix} 0 & 1 \\ -1 & 0 \end{bmatrix}$ is known as the *symplectic form*. Note that with this matrix notation the position–momentum commutators read $[\hat{R}_j, \hat{R}_l] = 2i\Omega_{jl}$. We will learn more about the physical properties of the displacement operator in the following sections, but for now, just take it as a useful mathematical object.

This operator can be written as a concatenation of two individual momentum and position translations plus some phase[3],

$$\hat{D}(\mathbf{r}) = \exp\left[-\frac{i}{4}px\right]\exp\left[\frac{i}{2}p\hat{X}\right]\exp\left[-\frac{i}{2}x\hat{P}\right], \tag{4.38}$$

which allows us to write

$$\text{tr}\left\{\hat{D}(\mathbf{r})\right\} = e^{-ipx/4}\int_{\mathbb{R}^2} dx' dp' \langle x'|e^{ip\hat{X}/2}e^{-ix\hat{P}/2}|p'\rangle\langle p'|x'\rangle \tag{4.39}$$

$$= 4\pi e^{-ipx/4}\underbrace{\int_{\mathbb{R}} \frac{dx'}{4\pi}e^{ipx'/2}}_{\delta(p)}\underbrace{\int_{\mathbb{R}} \frac{dp'}{4\pi}e^{ixp'/2}}_{\delta(x)}, \tag{4.40}$$

arriving at the identity

$$\text{tr}\left\{\hat{D}(\mathbf{r})\right\} = 4\pi\delta^{(2)}(\mathbf{r}), \tag{4.41}$$

which will be useful in many situations. Another important property, trivially proved from (4.37), is

$$\hat{D}(\mathbf{r})\hat{D}(\mathbf{r}') = \exp\left(-\frac{i}{4}\mathbf{r}^T \Omega \mathbf{r}'\right)\hat{D}(\mathbf{r} + \mathbf{r}'), \tag{4.42}$$

and therefore, except for a phase, the concatenation of two displacement operators is equivalent to a single displacement operator with the sum of the displacements.

---

[3] This is trivially proved by using the so-called disentangling lemma

$$\exp\left(\hat{A} + \hat{B}\right) = \exp\left(-[\hat{A}, \hat{B}]/2\right)\exp\left(\hat{A}\right)\exp\left(\hat{B}\right), \tag{4.37}$$

valid for operators $\hat{A}$ and $\hat{B}$ which commute with their commutator.

Note that the phase is zero only when the condition $xp' = px'$ is met, although it plays no physical role when applied to a state of the system.

The first step in order to find the Wigner function of a given state $\hat{\rho}$ from the displacement operator is to define the *characteristic function*

$$\chi_{\hat{\rho}}(\mathbf{r}) = \text{tr}\left\{\hat{D}(\mathbf{r})\hat{\rho}\right\} \iff \hat{\rho} = \int_{\mathbb{R}^2} \frac{d^2\mathbf{r}}{4\pi}\chi_{\hat{\rho}}(\mathbf{r})\hat{D}^\dagger(\mathbf{r}), \tag{4.43}$$

and then the Wigner function is obtained as its two-dimensional Fourier transform

$$W_{\hat{\rho}}(\mathbf{r}) = \int_{\mathbb{R}^2} \frac{d^2\mathbf{s}}{(4\pi)^2}\chi_{\hat{\rho}}(\mathbf{s})e^{\frac{i}{2}\mathbf{s}^T\Omega\mathbf{r}} \iff \chi_{\hat{\rho}}(\mathbf{s}) = \int_{\mathbb{R}^2} d^2\mathbf{r}\, W_{\hat{\rho}}(\mathbf{r})e^{-\frac{i}{2}\mathbf{s}^T\Omega\mathbf{r}}. \tag{4.44}$$

It is not difficult to show that this alternative definition of the Wigner function leads to the original one given by (4.32). However, this definition simplifies many more derivations. For example, from (4.43) and (4.44), it is immediate to prove that

$$\text{tr}\{\hat{\rho}\} = 1 \implies \chi_{\hat{\rho}}(0) = 1 \implies \int_{\mathbb{R}^2} d^2\mathbf{r}\, W_{\hat{\rho}}(\mathbf{r}) = 1, \tag{4.45}$$

that is, the normalization of the Wigner function. Moreover, evaluating the trace of (4.43) in the position eigenbasis, we obtain

$$\chi_{\hat{\rho}}(0, p) = \int dx \langle x|e^{ip\hat{X}/2}\hat{\rho}|x\rangle = \int dx e^{ipx/2}\langle x|\hat{\rho}|x\rangle, \tag{4.46}$$

which, using the right-hand side of (4.44) directly implies that

$$\langle x|\hat{\rho}|x\rangle = \int_{\mathbb{R}} dp\, W_{\hat{\rho}}(x, p), \tag{4.47}$$

that is, the Wigner function has the position probability density function as one of its marginals. The other marginal is the momentum probability density function, as is proved from $\chi_{\hat{\rho}}(x, 0)$ in a similar fashion.

Going one step further, using these definitions it is actually quite simple to prove (4.33). In particular, it is easy to check from (4.43) that the expectation value of the symmetrically ordered product $(\hat{X}^m\hat{P}^n)^{(s)}$ can be obtained from the characteristic function as

$$\left\langle \left(\hat{X}^m\hat{P}^n\right)^{(s)} \right\rangle = (-1)^m(2i)^{m+n}\frac{\partial^{m+n}}{\partial p^m \partial x^n}\chi_{\hat{\rho}}(\mathbf{r})\bigg|_{\mathbf{r}=0}, \tag{4.48}$$

leading to (4.33) after making use of (4.43).

4.3.3 Multi-mode considerations

In general we will not deal with a single harmonic oscillator (a single mode of the light field), but with a collection of, say, N harmonic oscillators (N modes of light). Let us define again the coordinate vector in the whole phase space as $\mathbf{r} = (x_1, p_1, x_2, p_2, ..., x_N, p_N)$, and the corresponding vector operator

$$\hat{\mathbf{R}} = \text{col}\left(\hat{X}_1, \hat{P}_1, \hat{X}_2, \hat{P}_2, ..., \hat{X}_N, \hat{P}_N\right), \tag{4.49}$$

in terms of which the commutation relations can be rewritten as

$$\left[\hat{R}_j, \hat{R}_l \right] = 2\mathrm{i}(\Omega_N)_{jl}, \qquad (4.50)$$

where

$$\Omega_N = \bigoplus_{m=1}^{N} \Omega = \begin{bmatrix} \Omega & & & \\ & \Omega & & \\ & & \ddots & \\ & & & \Omega \end{bmatrix} \left(= -\Omega_N^T = -\Omega_N^{-1}\right), \qquad (4.51)$$

is the symplectic form of N modes (in the following we will suppress the subindex N except when needed). Note that the modes are labeled here by numbers, but we can use any other suitable label; for example, when working with two modes shared by Alice and Bob it is natural to denote them by A and B, or in a system–environment scenario by S and E.

The state of the system acts now onto the tensor product of the Hilbert spaces of the modes, and so does the displacement operator, which is now defined as

$$\hat{D}(\mathbf{r}) = \exp\left[\frac{\mathrm{i}}{2} \hat{\mathbf{R}}^T \Omega \mathbf{r} \right] = \hat{D}(\mathbf{r}_1) \otimes \hat{D}(\mathbf{r}_2) \otimes \ldots \otimes \hat{D}(\mathbf{r}_N), \qquad (4.52)$$

and satisfies[4] $\mathrm{tr}\{\hat{D}(\mathbf{r})\} = (4\pi)^N \delta^{(2N)}(\mathbf{r})$, as well as (4.42). The characteristic function is defined as before

$$\chi_{\hat{\rho}}(\mathbf{r}) = \mathrm{tr}\left\{ \hat{D}(\mathbf{r})\hat{\rho} \right\} \iff \hat{\rho} = \int_{\mathbb{R}^{2N}} \frac{\mathrm{d}^{2N}\mathbf{r}}{(4\pi)^N} \chi_{\hat{\rho}}(\mathbf{r})\hat{D}^\dagger(\mathbf{r}), \qquad (4.54)$$

and the Wigner function as its $2N$-dimensional Fourier transform

$$W_{\hat{\rho}}(\mathbf{r}) = \int_{\mathbb{R}^{2N}} \frac{\mathrm{d}^{2N}\mathbf{s}}{(4\pi)^{2N}} \chi_{\hat{\rho}}(\mathbf{s}) e^{\frac{\mathrm{i}}{2}\mathbf{s}^T \Omega \mathbf{r}} \iff \chi_{\hat{\rho}}(\mathbf{s}) = \int_{\mathbb{R}^{2N}} \mathrm{d}^{2N}\mathbf{r}\, W_{\hat{\rho}}(\mathbf{r}) e^{-\frac{\mathrm{i}}{2}\mathbf{s}^T \Omega \mathbf{r}}. \quad (4.55)$$

Symmetric multi-mode moments can be evaluated from the characteristic or Wigner functions similarly to how we did in the single-mode case,

$$\left\langle \left(\hat{X}_1^{m_1} \hat{P}_1^{n_1} \right)^{(s)} \ldots \left(\hat{X}_N^{m_N} \hat{P}_N^{n_N} \right)^{(s)} \right\rangle = \int_{\mathbb{R}^{2N}} \mathrm{d}^{2N}\mathbf{r}\, W_{\hat{\rho}}(\mathbf{r}) x_1^{m_1} p_1^{n_1} \ldots x_N^{m_N} p_N^{n_N} \qquad (4.56a)$$

$$= (-1)^{m_1 + \cdots + m_N} (2\mathrm{i})^{m_1 + n_1 + \cdots + m_N + n_N} \frac{\partial^{m_1 + n_1 + \cdots + m_N + n_N}}{\partial p_1^{m_1} \partial x_1^{n_1} \ldots \partial p_N^{m_N} \partial x_N^{n_N}} \chi_{\hat{\rho}}(\mathbf{r})\bigg|_{\mathbf{r}=0}. \qquad (4.56b)$$

---

[4] Note that the $2N$-dimensional Dirac delta function can be written as

$$\delta^{(2N)}(\mathbf{r}) = \int_{\mathbb{R}^{2N}} \frac{\mathrm{d}^{2N}\mathbf{s}}{(4\pi)^{2N}} e^{\frac{\mathrm{i}}{2}\mathbf{s}^T \Omega \mathbf{r}}. \qquad (4.53)$$

In the previous chapters, we saw that there is an operation that plays an important role when dealing with composite Hilbert spaces: the partial trace. For example, in the case of the N harmonic oscillators being in a state $\hat{\rho}$, imagine that we want to trace out the last one, obtaining the reduced state of the remaining $N - 1$ oscillators, $\hat{\rho}_R = \mathrm{tr}_N\{\hat{\rho}\}$; let us see what this means in phase space. From (4.43) we see that the characteristic function of the reduced state $\hat{\rho}_R$ is just the original one with the phase-space coordinates of the traced oscillator set to zero, that is,

$$\chi_R(\mathbf{r}_{\{N-1\}}) = \mathrm{tr}_{\{N-1\}}\left\{\hat{D}(\mathbf{r}_{\{N-1\}})\hat{\rho}_R\right\} = \mathrm{tr}\left\{\hat{D}(\mathbf{r}_{\{N-1\}})\hat{\rho}\right\} = \chi_{\hat{\rho}}\left(\mathbf{r}_{\{N-1\}}, \mathbf{0}\right), \qquad (4.57)$$

where we use the notation $\mathbf{r}_{\{N-1\}} = \mathrm{col}(\mathbf{r}_1, \mathbf{r}_2, ..., \mathbf{r}_{N-1})$. Therefore, the Wigner function associated with the reduced state $\hat{\rho}_R$ can be found by integrating out the phase-space variables of the corresponding oscillator, that is,

$$W_R(\mathbf{r}_{\{N-1\}}) = \int_{\mathbb{R}^{2(N-1)}} \frac{\mathrm{d}^{2(N-1)}\mathbf{s}_{\{N-1\}}}{(4\pi)^{2(N-1)}}\chi_R(\mathbf{s}_{\{N-1\}})\mathrm{e}^{\frac{\mathrm{i}}{2}\mathbf{s}_{\{N-1\}}^T \Omega_{N-1}\mathbf{r}_{\{N-1\}}}$$

$$= \int_{\mathbb{R}^2} \mathrm{d}^2\mathbf{r}_N \, W_{\hat{\rho}}(\mathbf{r}_{\{N-1\}}, \mathbf{r}_N). \qquad (4.58)$$

4.4 Gaussian continuous-variable systems

4.4.1 Gaussian states

General definition
A particularly important class of quantum-mechanical states of the harmonic oscillator are the so-called *Gaussian states*, that is, states which have a Gaussian Wigner function. As we will see, this is the type of states which are most naturally generated in the laboratory, although we will also show how to design experimental schemes whose purpose is the generation of non-Gaussian states.

The Wigner function of an arbitrary single-mode Gaussian state has the form[5]

$$W(\mathbf{r}) = \frac{1}{2\pi\sqrt{\det V}} \exp\left[-\frac{1}{2}(\mathbf{r} - \bar{\mathbf{r}})^T V^{-1}(\mathbf{r} - \bar{\mathbf{r}})\right], \qquad (4.60)$$

where we have defined the *mean vector*

$$\bar{\mathbf{r}} = \left\langle \hat{\mathbf{R}} \right\rangle = \mathrm{col}\left(\left\langle \hat{X} \right\rangle, \left\langle \hat{P} \right\rangle\right), \qquad (4.61)$$

---

[5] When dealing with Gaussian states, the following integral is quite useful:

$$\int_{\mathbb{R}^{2N}} \mathrm{d}^{2N}\mathbf{r} \exp\left(-\frac{1}{2}\mathbf{r}^T A\mathbf{r} + \mathbf{x}^T\mathbf{r}\right) = \frac{(2\pi)^N}{\sqrt{\det A}} \exp\left(\frac{1}{2}\mathbf{x}^T A^{-1}\mathbf{x}\right), \qquad (4.59)$$

where $\mathbf{r} \in \mathbb{R}^{2N}$ and A is a non-singular $2N \times 2N$ matrix.

and the *covariance matrix*

$$V = \begin{bmatrix} \langle \delta \hat{X}^2 \rangle & \frac{1}{2} \langle \{ \delta \hat{X}, \delta \hat{P} \} \rangle \\ \frac{1}{2} \langle \{ \delta \hat{X}, \delta \hat{P} \} \rangle & \langle \delta \hat{P}^2 \rangle \end{bmatrix}, \tag{4.62}$$

whose elements are given by

$$V_{jl} = \frac{1}{2} \langle \{ \delta \hat{R}_j, \delta \hat{R}_l \} \rangle. \tag{4.63}$$

In this expression we have used the fluctuation operator $\delta \hat{A} = \hat{A} - \langle \hat{A} \rangle$, and denoted the anticommutator by curly brackets, $\{ \hat{A}, \hat{B} \} = \hat{A}\hat{B} + \hat{B}\hat{A}$. Note that number states do not belong to this class of states (save the vacuum, as we show below), but we will show that many other interesting states do. Note also that Gaussian states are completely defined by their first and second moments, which is why we will denote by $\hat{\rho}_G(\bar{\mathbf{r}}, V)$ a given Gaussian state.

Even though the covariance matrix appears here simply as a mathematical object with which we can write the Wigner function in a compact way, we will see that it has a huge physical significance. Indeed, note that Gaussian states have a positive Wigner function, and hence, the intuition of quantum mechanics corresponding to noise on phase space applies to them. As will be clear in the next sections, the covariance matrix contains precisely the way in which this noise is distributed in phase space, that is, it contains all the information about *quantum fluctuations*.

If the the world was classical, any covariance matrix would be allowed, as long as it was real, symmetric, and positive semidefinite. A quantum-mechanical harmonic oscillator has the added constraint $\det\{V\} \geqslant 1$, which comes from the uncertainty principle between position and momentum. Indeed, the following proof is quite reminiscent of the standard proof of the uncertainty principle:

$$\det\{V\} = \langle \delta \hat{X}^2 \rangle \langle \delta \hat{P}^2 \rangle - \frac{1}{4} \langle \{ \delta \hat{X}, \delta \hat{P} \} \rangle^2 \geqslant \left| \langle \delta \hat{X} \delta \hat{P} \rangle \right|^2 - \frac{1}{4} \langle \{ \delta \hat{X}, \delta \hat{P} \} \rangle^2$$

$$= \frac{1}{4} \left| \underbrace{\langle [\delta \hat{X}, \delta \hat{P}] \rangle}_{\text{imaginary}} + \underbrace{\langle \{ \delta \hat{X}, \delta \hat{P} \} \rangle}_{\text{real}} \right|^2 - \frac{1}{4} \langle \{ \delta \hat{X}, \delta \hat{P} \} \rangle^2 = 1, \tag{4.64}$$

where the inequality comes from the Cauchy–Schwarz inequality. A Gaussian state corresponding to any real, symmetric, positive-semidefinite covariance matrix satisfying this condition is physically achievable.

In the case of Gaussian states of N modes, their Wigner function takes the following form in the whole phase space:

$$W(\mathbf{r}) = \frac{1}{(2\pi)^N \sqrt{\det V}} \exp\left[-\frac{1}{2}(\mathbf{r} - \bar{\mathbf{r}})^T V^{-1}(\mathbf{r} - \bar{\mathbf{r}}) \right], \tag{4.65}$$

where now $\bar{\mathbf{r}}$ is a vector with $2N$ components $\bar{r}_j = \langle \hat{R}_j \rangle$, and V is a $2N \times 2N$ matrix with elements $V_{jl} = \langle \{\delta\hat{R}_j, \delta\hat{R}_l\} \rangle/2$. It is interesting to note that the mean photon number can be written as

$$\sum_{j=1}^{N} \langle \hat{N}_j \rangle = \frac{1}{4}\mathrm{tr}\{V\} + \frac{1}{4}\bar{\mathbf{r}}^2 - \frac{N}{2}. \tag{4.66}$$

In this case, the necessary and sufficient condition for a real, symmetric $2N \times 2N$ matrix V to correspond to a physical quantum state of N oscillators is [2]

$$V + i\Omega \geqslant 0, \tag{4.67}$$

which is again linked to the uncertainty principle, and implies the positivity of V. We will prove in section 4.4.3 that any Gaussian state for which the eigenvalues of $V + i\Omega$ are zero (equivalently, which saturates the uncertainty principle) is pure.

In many situations it is interesting to understand the state of the N oscillators as a bipartite state of M oscillators plus another M' oscillators ($M + M' = N$), in which case we talk about an $M \times M'$ continuous-variable system. Consider a Gaussian state $\hat{\rho}_G(\bar{\mathbf{r}}, V)$ of the N oscillators, whose mean vector and covariance matrix we write as

$$\bar{\mathbf{r}} = \mathrm{col}(\bar{\mathbf{x}}, \bar{\mathbf{x}}') \qquad \text{and} \qquad V = \begin{bmatrix} W & C \\ C^T & W' \end{bmatrix}, \tag{4.68}$$

where $\bar{\mathbf{x}} \in \mathbb{R}^{2M}$, $\bar{\mathbf{x}}' \in \mathbb{R}^{2M'}$, W and W' are real, symmetric matrices of dimensions $2M \times 2M$ and $2M' \times 2M'$, respectively, and C is a real matrix of dimensions $2M \times 2M'$. Then, it is easy to prove that the state of the first M modes after tracing out the remaining M' modes, is the Gaussian state

$$\hat{\rho}_G(\bar{\mathbf{x}}, W) = \mathrm{tr}_{M+1,M+2,\dots,M+M'}\{\hat{\rho}_G(\bar{\mathbf{r}}, V)\}, \tag{4.69}$$

that is, tracing out a mode in a Gaussian state is equivalent to removing its corresponding entries in the mean vector, as well as its rows and columns in the covariance matrix. In order to prove this we make use of the characteristic function, which for the general Gaussian Wigner function (4.65) takes the form

$$\chi(\mathbf{s}) = \exp\left[-\frac{1}{8}\mathbf{s}^T\Omega V\Omega^T\mathbf{s} + \frac{i}{4}\bar{\mathbf{r}}^T\Omega\mathbf{s} \right]. \tag{4.70}$$

Then, by substituting (4.68) in this expression, and remembering that tracing out any mode is equivalent to setting to zero the corresponding phase-space variables in the characteristic function, we can write the characteristic function of the reduced state of the first M modes as

$$\chi_R(\mathbf{s}_{\{M\}}) = \exp\left[-\frac{1}{8}\mathbf{s}_{\{M\}}^T\Omega_M W\Omega_M\mathbf{s}_{\{M\}} + \frac{i}{4}\bar{\mathbf{x}}^T\Omega_M\mathbf{s}_{\{M\}} \right], \tag{4.71}$$

which proves (4.69).

Finally, note that when the states of the partitions are uncorrelated, that is,

$$\hat{\rho}_G(\bar{\mathbf{r}}, V) = \hat{\rho}_G(\bar{\mathbf{x}}, W) \otimes \hat{\rho}_G(\bar{\mathbf{x}}', W'), \qquad (4.72)$$

the covariance matrix of the Wigner function (4.65) can be written as a direct sum

$$V = W \oplus W'. \qquad (4.73)$$

Examples of Gaussian states

The vacuum state. As commented above, number states are not Gaussian, with one exception: the *vacuum state* $|0\rangle$. To see this, just note that its Wigner function (4.34) can be written as

$$W_{|0\rangle}(x, p) = \frac{1}{2\pi} \exp\left(-\frac{x^2 + p^2}{2}\right), \qquad (4.74)$$

that is, a Gaussian distribution such as (4.60) with

$$\bar{\mathbf{r}} = \mathbf{0} \qquad \text{and} \qquad V = \begin{bmatrix} 1 & 0 \\ 0 & 1 \end{bmatrix} \equiv I_{2\times 2}. \qquad (4.75)$$

Hence, in Gaussian notation $|0\rangle\langle 0| = \hat{\rho}_G(\mathbf{0}, I_{2\times 2})$.

It is interesting to note that the vacuum state is a *minimum-uncertainty state*, that is, it carries the minimum amount of noise allowed by the uncertainty relations, $\Delta X \Delta P = 1$. Moreover, when the oscillator is in the vacuum state, its quantum fluctuations are equally distributed among position and momentum, $\Delta X = \Delta P = 1$. Indeed, this minimal quantum noise (commonly denoted by *shot noise* in quantum optics) is distributed homogeneously along all directions of phase space, as can be appreciated in figure 4.2.

Thermal states. As explained in sections 1.3.8 and 2.1, the mixedness of the state of a system can be related to the amount of information which has been lost to another system inaccessible to us, that is, to the correlations shared with this second system. Given a system whose associated Hilbert space has dimension d, and is spanned by some orthonormal basis $\{|j\rangle\}_{j=1,2,\ldots,d}$, we have already seen that its maximally mixed state is

$$\hat{\rho}_{MM} = \frac{1}{d} \sum_{j=1}^{d} |j\rangle\langle j| = \frac{1}{d} \hat{I}. \qquad (4.76)$$

As it is proportional to the identity, this state is invariant under changes of basis and, hence, the eigenvalues of any observable of the system are equally likely. This is in concordance with what one expects intuitively from a state which has leaked the maximum amount of information to another system.

For infinite-dimensional Hilbert spaces ($d \to \infty$) this state is not physical since it has infinite excitations, that is, $\text{tr}\{\hat{\rho}\hat{N}\} \to \infty$. Hence, finding the maximally mixed state in infinite dimensions makes sense only if one adds an 'energy' constraint such as $\mathrm{tr}\{\hat{\rho}\hat{N}\} = \bar{n}$, where $\bar{n}$ is positive real. It is possible to show that the state which maximizes the von Neumann entropy subject to this constraint is

$$\hat{\rho}_{\mathrm{th}}(\bar{n}) = \sum_{n=0}^{\infty} \frac{\bar{n}^n}{(1+\bar{n})^{1+n}} |n\rangle\langle n|, \tag{4.77}$$

which is still diagonal in the Fock basis, but does not have a flat distribution for the photon number. Interestingly, given the free Hamiltonian of the oscillator $\hat{H} = \hbar\omega(\hat{a}^{\dagger}\hat{a} + 1/2)$, the state can alternatively be written as

$$\hat{\rho}_{\mathrm{th}}(\bar{n}) = \frac{\exp(-\beta\hat{H})}{\mathrm{tr}\{\exp(-\beta\hat{H})\}}, \tag{4.78}$$

which, with the identification $\bar{n} = 1/(e^{\beta\hbar\omega} - 1)$, corresponds to the state expected for bosons at thermal equilibrium at temperature $T = 1/k_B\beta$, where k_B is the Boltzmann constant. This is why this state is known as the *thermal state*, whose von Neumann entropy reads

$$S[\hat{\rho}_{\mathrm{th}}(\bar{n})] = (\bar{n}+1)\log(\bar{n}+1) - \bar{n}\log\bar{n} \equiv S_{\mathrm{th}}(\bar{n}), \tag{4.79}$$

as is easily proven given that the state is diagonal in the Fock basis.

It is not difficult to see that this state is Gaussian (later on we will actually prove it by simple means), and that it is defined by a zero mean vector, and a covariance matrix

$$V_{\mathrm{th}}(\bar{n}) = (2\bar{n}+1)I_{2\times2}, \tag{4.80}$$

that is, $\hat{\rho}_{\mathrm{th}}(\bar{n}) = \hat{\rho}_{\mathrm{G}}[\mathbf{0}, V_{\mathrm{th}}(\bar{n})]$. The corresponding Wigner function can be seen in figure 4.3, where it can be appreciated that a thermal state is similar to a vacuum state, but with more noise. Consequently, the vacuum state can be seen as a thermal state with zero mean photon number.

In the next section we will learn that any N-mode Gaussian state can be decomposed into N uncorrelated thermal states (Williamson's theorem), and hence thermal states can be seen as the most fundamental Gaussian states.

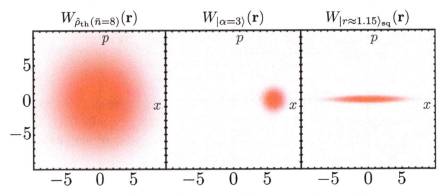

Figure 4.3. Density plot of the Wigner functions corresponding to a thermal state $\hat{\rho}_{\mathrm{th}}(\bar{n})$, a coherent state $|\alpha\rangle$, and a squeezed state $|r\rangle_{\mathrm{sq}}$.

4.4.2 Gaussian unitaries

General definition

Consider a unitary transformation $\hat{U} = \exp(-i\hat{H})$ acting on the state of the oscillator. We say that such a unitary is Gaussian when it maps Gaussian states into Gaussian states.

Let us define the vector operator $\hat{\mathbf{a}} = \mathrm{col}(\hat{a}_1, \hat{a}_2, ...,\hat{a}_N)$. It is quite intuitive that any Gaussian unitary will come from a Hamiltonian having only linear or bilinear terms, that is,

$$\hat{H} = \hat{\mathbf{a}}^\dagger \boldsymbol{\alpha} + \hat{\mathbf{a}}^\dagger \mathcal{F} \hat{\mathbf{a}} + \hat{\mathbf{a}}^\dagger \mathcal{G} \hat{\mathbf{a}}^{\dagger T} + \mathrm{H.\,c.}, \tag{4.81}$$

for some vector $\boldsymbol{\alpha} \in \mathbb{C}^N$, and some symmetric, complex $N \times N$ matrices $\mathcal{F}$ and $\mathcal{G}$. Let us try to understand why this is so by analyzing the physical meaning of each term. The first term corresponds to the injection of excitations in the modes 'one by one'. The second term comprises all the energy shifts $\hat{a}_j^\dagger \hat{a}_j$, as well as the creation of a photon in one mode via the annihilation of a photon of another mode, $\hat{a}_j^\dagger \hat{a}_l$, that is, a 'photon-by-photon' exchange. The last term takes into account the possibility of generating two photons simultaneously, $\hat{a}_j^\dagger \hat{a}_l^\dagger$. In other words, all the possible one-photon and two-photon processes are taken into account in this Hamiltonian. It is intuitive that if we want to transform Gaussian states into Gaussian states, the Hamiltonian cannot contain multi-photon processes beyond these, because otherwise one would create correlations which go beyond the ones captured by first and second moments.

In the Heisenberg picture the Gaussian unitary induces then a linear transformation (known as *Bogoliubov transformation*) of the type

$$\hat{\mathbf{a}} \longrightarrow \hat{U}^\dagger \hat{\mathbf{a}} \hat{U} = \mathcal{A} \hat{\mathbf{a}} + \hat{\mathbf{a}}^\dagger \mathcal{B} + \boldsymbol{\alpha}, \tag{4.82}$$

where the form of the complex $N \times N$ matrices $\mathcal{A}$ and $\mathcal{B}$ in terms of $(\boldsymbol{\alpha}, \mathcal{F}, \mathcal{G})$ is unimportant for our purposes. The only restriction on these is that they have to satisfy $\mathcal{A}\mathcal{B} = \mathcal{B}^T \mathcal{A}^T$ and $\mathcal{A}\mathcal{A}^\dagger = \mathcal{B}^\dagger \mathcal{B} + \mathcal{I}_{N \times N}$, in order to preserve the commutation relations of the creation and annihilation operators ($\mathcal{I}_{N \times N}$ is the identity matrix of dimension N).

Instead of writing this linear transformation in terms of the bosonic operators, one can write it in terms of the position and momentum, or, more compactly, in terms of $\hat{\mathbf{R}}$:

$$\hat{\mathbf{R}} \longrightarrow \hat{U}^\dagger \hat{\mathbf{R}} \hat{U} = S\hat{\mathbf{R}} + \mathbf{d}, \tag{4.83}$$

where, once again, the dependence of $\mathbf{d} \in \mathbb{R}^{2N}$ and the real $2N \times 2N$ matrix S in the previous transformation parameters is unimportant for our purposes. The only relevant issue is that, in order to preserve the commutation relations of the quadratures, S has to satisfy

$$S\Omega S^T = \Omega, \tag{4.84}$$

that is, S must be a *symplectic matrix* [3]. In the following we will denote Gaussian unitaries by $\hat{U}_{\mathrm{G}}(\mathbf{d}, S)$ to stress the fact that they are completely characterized by $\mathbf{d}$ and S.

The transformation induced onto the system by the Gaussian unitary $\hat{U}_G(\mathbf{d}, S)$ is easily described in the Schrödinger picture as well if the states are represented by the Wigner function. To see this, let us first note that the displacement operator is transformed by the action of this unitary as

$$\hat{U}_G^\dagger \hat{D}(\mathbf{r})\hat{U}_G = \exp\left[\frac{i}{2}\left(S\hat{\mathbf{R}} + \mathbf{d}\right)^T \Omega\mathbf{r}\right]$$

$$\underset{S^T\Omega=\Omega S^{-1}}{\equiv} \exp\left[\frac{i}{2}\left(\hat{\mathbf{R}}^T\Omega S^{-1} + \mathbf{d}^T\Omega\right)\mathbf{r}\right] = \hat{D}\left(S^{-1}\mathbf{r}\right)\exp\left[\frac{i}{2}\mathbf{d}^T\Omega\mathbf{r}\right], \quad (4.85)$$

where the identity $S^T\Omega = \Omega S^{-1}$ follows from (4.84) and (4.51). Therefore, given the initial state $\hat{\rho}$ of the system, the corresponding characteristic function is transformed as

$$\chi_{\hat{U}_G\hat{\rho}\hat{U}_G^\dagger}(\mathbf{r}) = \mathrm{tr}\left\{\hat{D}(\mathbf{r})\hat{U}_G\hat{\rho}\,\hat{U}_G^\dagger\right\} = e^{\frac{i}{2}\mathbf{d}^T\Omega\mathbf{r}}\mathrm{tr}\left\{\hat{D}\left(S^{-1}\mathbf{r}\right)\hat{\rho}\right\} = e^{-\frac{i}{2}\mathbf{r}^T\Omega\mathbf{d}}\chi_{\hat{\rho}}(S^{-1}\mathbf{r}), \quad (4.86)$$

and the Wigner function as

$$W_{\hat{U}_G\hat{\rho}\hat{U}_G^\dagger}(\mathbf{r}) = \int_{\mathbb{R}^{2N}} \frac{d^{2N}\mathbf{s}}{(4\pi)^{2N}}\chi_{\hat{\rho}}\left(S^{-1}\mathbf{s}\right)e^{\frac{i}{2}\mathbf{s}^T\Omega(\mathbf{r}-\mathbf{d})} \underset{\mathbf{z}=S^{-1}\mathbf{s}}{\equiv} \int_{\mathbb{R}^{2N}} \frac{d^{2N}\mathbf{z}}{(4\pi)^{2N}}\chi_{\hat{\rho}}(\mathbf{z})e^{\frac{i}{2}\mathbf{z}^T S^T\Omega(\mathbf{r}-\mathbf{d})}$$

$$\underset{S^T\Omega=\Omega S^{-1}}{\equiv} W_{\hat{\rho}}[S^{-1}(\mathbf{r} - \mathbf{d})]. \quad (4.87)$$

In the case of Gaussian states, the situation is even simpler: one only needs to find the effect of the transformation onto the first and second moments of the state. Using (4.83), it is straightforward to show that the transformation induced by the Gaussian unitary $\hat{U}_G(\mathbf{d}, S)$ on the mean vector $\bar{\mathbf{r}}$ and the covariance matrix V of any state is

$$\bar{\mathbf{r}} \longrightarrow S\bar{\mathbf{r}} + \mathbf{d} \qquad \text{and} \qquad V \longrightarrow SVS^T. \quad (4.88)$$

Taking the $M \times M'$ partition of the multi-mode system as in the previous section, note that when the unitary transformation acts independently on each partition, that is,

$$\hat{U}_G(\mathbf{d}, S) = \hat{U}_G(\mathbf{l}, \mathcal{L}) \otimes \hat{U}_G(\mathbf{l}', \mathcal{L}'), \quad (4.89)$$

where $\mathbf{l} \in \mathbb{R}^{2M}$, $\mathbf{l}' \in \mathbb{R}^{2M'}$, and $\mathcal{L}$ and $\mathcal{L}'$ are symplectic matrices of dimensions $2M \times 2M$ and $2M' \times 2M'$, respectively, we can write

$$\mathbf{d} = (\mathbf{l}, \mathbf{l}') \qquad \text{and} \qquad S = \mathcal{L} \oplus \mathcal{L}'. \quad (4.90)$$

Finally, its important to remark that a Gaussian unitary transformation is called *passive* when it conserves the mean photon number $\sum_{j=1}^N \langle\hat{N}_j\rangle$ of Gaussian states, and *active* if it changes it. Now, given the transformation (4.88), and keeping in mind that the mean photon number is proportional to the square of the mean vector, $\bar{\mathbf{r}}^2$, and the trace of the covariance matrix, $\mathrm{tr}\{V\}$, see (4.66), a Gaussian unitary will be passive if and only if

$$\mathbf{d} = 0 \qquad \text{and} \qquad S^T S = I_{2N \times 2N}, \tag{4.91}$$

the second condition meaning that its associated symplectic transformation must be *orthogonal*, that is, $S^T = S^{-1}$.

Examples of Gaussian unitaries and more Gaussian states
The displacement operator and coherent states. Consider the unitary operator

$$\hat{D}(\alpha) = \exp(\alpha \hat{a}^\dagger - \alpha^* \hat{a}), \tag{4.92}$$

which, using the formula (4.37) can be written in the equivalent way

$$\hat{D}(\alpha) = \exp(-|\alpha|^2/2) \exp(\alpha \hat{a}^\dagger) \exp(-\alpha^* \hat{a}), \tag{4.93}$$

which we will refer to as its *normal form*.

Using the Baker–Campbell–Haussdorf lemma (4.16), it is fairly simple to prove that this operator transforms the annihilation operator as

$$\hat{a} \to \hat{D}^\dagger(\alpha)\hat{a}\hat{D}(\alpha) = \hat{a} + \alpha, \tag{4.94}$$

or, in terms of the quadratures

$$\hat{X} \to \hat{D}^\dagger(\alpha)\hat{X}\hat{D}(\alpha) = \hat{X} + x_\alpha, \tag{4.95a}$$

$$\hat{P} \to \hat{D}^\dagger(\alpha)\hat{P}\hat{D}(\alpha) = \hat{P} + p_\alpha, \tag{4.95b}$$

where $x_\alpha = \alpha^* + \alpha$ and $p_\alpha = i(\alpha^* - \alpha)$. This unitary operator is then called the *displacement operator* because it allows us to perform translations in phase space. Indeed, it is exactly the operator that we defined in the previous section, see (4.36), what is easily shown by rewriting (4.92) in terms of the position and momentum operators.

As a Gaussian unitary we then write $\hat{D}(\alpha) = \hat{U}_G(\mathbf{d}_\alpha, I_{2\times 2})$ with

$$\mathbf{d}_\alpha = \mathrm{col}(x_\alpha, p_\alpha). \tag{4.96}$$

The states obtained by displacing the vacuum state are known as *coherent states*. Using the normal form of the displacement operator, it is easy to obtain

$$|\alpha\rangle = \hat{D}(\alpha)|0\rangle = \sum_{n=0}^\infty \frac{\exp(-|\alpha|^2/2)\alpha^{n/2}}{\sqrt{n!}}|n\rangle. \tag{4.97}$$

These are Gaussian states with the same covariance matrix as the vacuum, but with a non-zero mean vector, that is,

$$\bar{\mathbf{r}} = \mathbf{d}_\alpha \qquad \text{and} \qquad V = I_{2\times 2}, \tag{4.98}$$

or, in Gaussian notation, $|\alpha\rangle\langle\alpha| = \hat{\rho}_G(\mathbf{d}_\alpha, I_{2\times 2})$. Hence, these states have the same noise properties as the vacuum; however, they describe a bright mode whose mean is shifted away from the origin of phase space, as shown in figure 4.3. In this sense, they can be interpreted as the 'most classical' states, since they carry the minimal amount of noise that quantum mechanics allows (the shot noise), equally distributed along all directions of phase space, with α playing the role of the (normalized) classical normal variable of the oscillator or the average field that would be measured in an experiment. In the case of light, they are a fair approximation to the state describing the beam coming out from a (phase-locked) laser.

From a mathematical point of view, they are the eigenstates of the annihilation operator, that is, $\hat{a}|\alpha\rangle = \alpha|\alpha\rangle$. Later on we will learn that, even though the annihilation operator is not self-adjoint, we can build a POVM-based measurement whose possible outcomes are the eigenvalues α (*heterodyne detection*).

The squeezing operator and squeezed states. Consider now the *squeezing operator*

$$\hat{S}(r) = \exp\left(\frac{r}{2}\hat{a}^{\dagger 2} - \frac{r}{2}\hat{a}^2\right), \tag{4.99}$$

where $r \in [0, \infty[$. This operator is implemented experimentally for an optical mode of frequency ω_0 by pumping a non-linear crystal with a strong laser beam of twice that frequency; pairs of photons at frequency ω_0 are generated via the so-called *spontaneous parametric down-conversion process* (see [4] and references therein).

Using the Baker–Campbell–Haussdorf lemma (4.16), it is again simple to prove that this operator transforms the annihilation operator as

$$\hat{a} \to \hat{S}^{\dagger}(r)\hat{a}\hat{S}(r) = \hat{a} \cosh r + \hat{a}^{\dagger} \sinh r, \tag{4.100}$$

or, in terms of the quadratures

$$\hat{X} \to \hat{S}^{\dagger}(r)\hat{X}\hat{S}(r) = e^{r}\hat{X}, \tag{4.101a}$$

$$\hat{P} \to \hat{S}^{\dagger}(r)\hat{P}\hat{S}(r) = e^{-r}\hat{P}, \tag{4.101b}$$

so that it is characterized as a Gaussian unitary by $\hat{S}(r) = \hat{U}_{\mathrm{G}}[\mathbf{0}, Q(r)]$ with

$$Q(r) = \begin{bmatrix} e^r & 0 \\ 0 & e^{-r} \end{bmatrix}. \tag{4.102}$$

The application of the squeezing operator to the vacuum state leads to a so-called *squeezed vacuum state*. In the Fock basis, this state is characterized by containing only even number states, which arises from the fact that the squeezing operator generates pairs of excitations. Its explicit representation in this basis is [5]

$$|r\rangle_{\mathrm{sq}} = \hat{S}(r)|0\rangle = \sum_{n=0}^{\infty} \frac{1}{2^n n!} \sqrt{\frac{(2n)!}{\cosh r}} \tanh^n r \, |2n\rangle. \tag{4.103}$$

This Gaussian state has a zero mean, and its covariance matrix is

$$V_{\mathrm{sq}}(r) = Q(r)Q^T(r) = \begin{bmatrix} e^{2r} & 0 \\ 0 & e^{-2r} \end{bmatrix}, \tag{4.104}$$

that is, $|r\rangle_{sq}\langle r| = \hat{\rho}_G[\mathbf{0}, V_{sq}(r)]$. This state is then a minimum-uncertainty state (since it saturates the uncertainty relation, $\Delta X \Delta P = 1$), which shows a reduced variance of the momentum with respect to the vacuum state, as shown in figure 4.3. Note that in the limit $r \rightarrow \infty$ the variance of the momentum goes to zero, while the variance of the position goes to infinity, and hence in the limit of infinite squeezing the state (4.103) is an eigenstate of the momentum operator. Note, however, that this limit is unphysical, as the mean photon number $\langle \hat{N} \rangle = \sinh^2 r$ diverges, and hence, an infinite amount of energy is required for the generation of a position or momentum eigenstate.

The phase-shift operator. The free evolution of an oscillator (corresponding to the free propagation of an optical mode through a linear medium) induces the unitary transformation

$$\hat{R}(\theta) = \exp(-i\theta\hat{a}^\dagger\hat{a}), \qquad (4.105)$$

known as the *phase-shift operator*, which transforms the annihilation operator as

$$\hat{a} \rightarrow \hat{R}^\dagger(\theta)\hat{a}\hat{R}(\theta) = \exp(i\theta)\hat{a}, \qquad (4.106)$$

or in terms of the position and momentum

$$\hat{X} \rightarrow \hat{R}^\dagger(\theta)\hat{X}\hat{R}(\theta) = \hat{X}\cos\theta + \hat{P}\sin\theta, \qquad (4.107a)$$

$$\hat{P} \rightarrow \hat{R}^\dagger(\theta)\hat{P}\hat{R}(\theta) = \hat{P}\cos\theta - \hat{X}\sin\theta. \qquad (4.107b)$$

Hence, as a Gaussian unitary this transformation is characterized by $\hat{R}(\theta) = \hat{U}_G[\mathbf{0}, \mathcal{R}(\theta)]$, where

$$\mathcal{R}(\theta) = \begin{bmatrix} \cos\theta & \sin\theta \\ -\sin\theta & \cos\theta \end{bmatrix}, \qquad (4.108)$$

which shows that a phase shift is equivalent to a *proper rotation* in phase space.

Note that number states are invariant under this transformation since they are eigenstates of $\hat{R}(\theta)$, and hence, thermal states are invariant under rotations in phase space. This is not the case for coherent or squeezed states, as is clear from figure 4.3.

The two-mode squeezing operator and two-mode squeezed states. All the unitaries considered so far act on a single mode, and hence, they cannot be used to induce entanglement between several modes. In this example we consider the *two-mode squeezing operator*

$$\hat{S}_{12}(r) = \exp\left(r\hat{a}_1^\dagger\hat{a}_2^\dagger - r\hat{a}_1\hat{a}_2\right), \qquad (4.109)$$

which can be implemented experimentally via a non-linear crystal similarly to the squeezing operator (4.99), but now in a regime in which the down-converted photons are distinguishable either in frequency, and/or polarization, and/or spatial mode [4].

Under the action of this operator, the annihilation operators transform as

$$\hat{a}_1 \rightarrow \hat{S}_{12}^\dagger(r)\hat{a}_1\hat{S}_{12}(r) = \hat{a}_1\cosh r + \hat{a}_2^\dagger\sinh r, \qquad (4.110a)$$

$$\hat{a}_2 \rightarrow \hat{S}_{12}^\dagger(r)\hat{a}_2\hat{S}_{12}(r) = \hat{a}_2 \cosh r + \hat{a}_1^\dagger \sinh r, \qquad (4.110b)$$

or, in terms of the quadratures

$$\hat{X}_1 \rightarrow \hat{S}_{12}^\dagger(r)\hat{X}_1\hat{S}_{12}(r) = \hat{X}_1 \cosh r + \hat{X}_2 \sinh r, \qquad (4.111a)$$

$$\hat{P}_1 \rightarrow \hat{S}_{12}^\dagger(r)\hat{P}_1\hat{S}_{12}(r) = \hat{P}_1 \cosh r - \hat{P}_2 \sinh r, \qquad (4.111b)$$

$$\hat{X}_2 \rightarrow \hat{S}_{12}^\dagger(r)\hat{X}_2\hat{S}_{12}(r) = \hat{X}_2 \cosh r + \hat{X}_1 \sinh r, \qquad (4.111c)$$

$$\hat{P}_2 \rightarrow \hat{S}_{12}^\dagger(r)\hat{P}_2\hat{S}_{12}(r) = \hat{P}_2 \cosh r - \hat{P}_1 \sinh r. \qquad (4.111d)$$

Hence, as a Gaussian unitary, this transformation is characterized by $\hat{S}_{12}(r) = \hat{U}_G[\mathbf{0}, Q_{12}(r)]$ with

$$Q_{12}(r) = \begin{bmatrix} I_{2\times2}\cosh r & \mathcal{Z}\sinh r \\ \mathcal{Z}\sinh r & I_{2\times2}\cosh r \end{bmatrix}, \qquad (4.112)$$

where $\mathcal{Z} = \mathrm{diag}(1, -1)$.

Applying the two-mode squeezing operator to a vacuum state, one obtains a so-called *two-mode squeezed vacuum state*. In the number state basis, this state is characterized by a perfectly correlated statistics of the number of quanta in the modes, which, once again, is a result of the fact that the two-mode squeezing operator generates pairs of excitations. Its explicit representation in this basis is [5]

$$|r\rangle_{2\mathrm{sq}} = \hat{S}_{12}(r)|0, 0\rangle = \frac{1}{\cosh r}\sum_{n=0}^{\infty} \tanh^n r|n, n\rangle, \qquad (4.113)$$

where we have used the notation $|n\rangle \otimes |m\rangle \equiv |n, m\rangle$. This Gaussian state has a zero mean, and its covariance matrix is

$$V_{2\mathrm{sq}}(r) = Q_{12}(r)Q_{12}^T(r) = \begin{bmatrix} I_{2\times2}\cosh 2r & \mathcal{Z}\sinh 2r \\ \mathcal{Z}\sinh 2r & I_{2\times2}\cosh 2r \end{bmatrix}, \qquad (4.114)$$

that is, $|r\rangle_{2\mathrm{sq}}\langle r| = \hat{\rho}[\mathbf{0}, V_{2\mathrm{sq}}(r)]$.

Note that by taking the partial trace with respect to any of its two modes, the two-mode squeezed vacuum state becomes a thermal state with mean photon number $\bar{n} = \sinh^2 r$, that is $\mathrm{tr}_2\{|r\rangle_{2\mathrm{sq}}\langle r|\} = \mathrm{tr}_1\{|r\rangle_{2\mathrm{sq}}\langle r|\} = \hat{\rho}_{\mathrm{th}}(\sinh^2 r)$, and hence the two-mode squeezed vacuum state can be seen as the purification of a thermal state, which in addition shows that it is the maximally entangled state in infinite dimensions for a fixed energy. We will come back to the entanglement properties of the two-mode squeezed vacuum state in section 4.4.4.

The beam-splitter operator. We are going to analyze only one more type of two-mode unitary transformations, the one induced by the so-called *beam-splitter operator*

$$\hat{B}_{12}(\beta) = \exp\left(\beta \hat{a}_1 \hat{a}_2^\dagger - \beta \hat{a}_1^\dagger \hat{a}_2\right), \tag{4.115a}$$

which can be implemented experimentally on light by, for example, mixing two optical beams in a beam splitter of *transmissivity* $T = \cos^2 \beta$.

Under the action of this operator, the annihilation operators transform as

$$\hat{a}_1 \rightarrow \hat{B}^\dagger(\beta)\hat{a}_1\hat{B}(\beta) = \hat{a}_1 \cos \beta + \hat{a}_2 \sin \beta \tag{4.116}$$

$$\hat{a}_2 \rightarrow \hat{B}^\dagger(\beta)\hat{a}_2\hat{B}(\beta) = \hat{a}_2 \cos \beta - \hat{a}_1 \sin \beta. \tag{4.117}$$

or, in terms of the quadratures

$$\hat{X}_1 \rightarrow \hat{B}_{12}^\dagger(\beta)\hat{X}_1\hat{B}_{12}(\beta) = \hat{X}_1 \cos \beta - \hat{X}_2 \sin \beta, \tag{4.118}$$

$$\hat{P}_1 \rightarrow \hat{B}_{12}^\dagger(\beta)\hat{P}_1\hat{B}_{12}(\beta) = \hat{P}_1 \cos \beta - \hat{P}_2 \sin \beta, \tag{4.119}$$

$$\hat{X}_2 \rightarrow \hat{B}_{12}^\dagger(\beta)\hat{X}_2\hat{B}_{12}(\beta) = \hat{X}_2 \cos \beta + \hat{X}_1 \sin \beta, \tag{4.120}$$

$$\hat{P}_2 \rightarrow \hat{B}_{12}^\dagger(\beta)\hat{P}_2\hat{B}_{12}(\beta) = \hat{P}_2 \cos \beta + \hat{P}_1 \sin \beta. \tag{4.121}$$

Hence, as a Gaussian unitary, this transformation is characterized by $\hat{B}_{12}(\beta) = \hat{U}_G[\mathbf{0}, \mathcal{B}_{12}(\beta)]$ with

$$\mathcal{B}_{12}(\beta) = \begin{bmatrix} \mathcal{I}_{2\times2} \cos \beta & -\mathcal{I}_{2\times2} \sin \beta \\ \mathcal{I}_{2\times2} \sin \beta & \mathcal{I}_{2\times2} \cosh \beta \end{bmatrix}. \tag{4.122}$$

Interestingly, note that when the states of both modes are coherent, they remain coherent after the action of the beam-splitter transformation, as $\mathcal{B}_{12}(\beta)\mathcal{B}_{12}^T(\beta) = \mathcal{I}_{4\times4}$. As an example, consider the state $|\alpha\rangle \otimes |0\rangle$ (one mode in an arbitrary coherent state, and the other in vacuum), which has the Gaussian representation $\hat{\rho}_G(\mathbf{d}, \mathcal{I}_{4\times4})$ with

$$\mathbf{d} = 2\mathrm{col}(\mathrm{Re}\{\alpha\}, \mathrm{Im}\{\alpha\}, 0, 0); \tag{4.123}$$

after the action of the beam-splitter operator, it becomes $\hat{\rho}_G'(\mathbf{d}', \mathcal{I}_{4\times4})$ with

$$\mathbf{d}' = 2 \, \mathrm{col}(\mathrm{Re}\{\alpha\}\cos \beta, \mathrm{Im}\{\alpha\}\cos \beta, \mathrm{Re}\{\alpha\}\sin \beta, \mathrm{Im}\{\alpha\}\sin \beta), \tag{4.124}$$

which is the tensor product of two coherent states, in particular, $|\alpha \cos \beta\rangle \otimes |\alpha \sin \beta\rangle$. This is exactly what one expects when a laser field is sent through a beam splitter: part of the laser is transmitted, and part is reflected.

4.4.3 General Gaussian unitaries and states

In this section we will use symplectic analysis (or better, 'symplectic tricks'), to find interesting facts about general Gaussian unitary transformations and Gaussian states.

Bloch–Messiah reduction: connection between single- and two-mode squeezed states, and the canonical form of general single-mode Gaussian states

It is well known in symplectic analysis [3, 6] that any $2N \times 2N$ symplectic matrix S can be decomposed as

$$S = \mathcal{K} \left[\bigoplus_{j=1}^{N} Q(r_j) \right] \mathcal{L}, \tag{4.125}$$

where $\mathcal{K}$ and $\mathcal{L}$ are orthogonal, symplectic matrices (this is known as the *Euler decomposition* of a symplectic matrix, or as its *Bloch–Messiah reduction*). Physically, this means that a general N-mode unitary transformation can be seen as the concatenation of three operations: an N-port interferometer mixing all the modes[6], N single-mode squeezers acting independently on each mode, and a second N-port interferometer.

As an important example involving two modes, note that the two-mode squeezing transformation can be written as

$$Q_{12}(r) = \mathcal{B}_{12}\left(-\frac{\pi}{4}\right)[Q(-r) \oplus Q(r)]\mathcal{B}_{12}\left(\frac{\pi}{4}\right), \tag{4.126}$$

which, in the Hilbert space, means that two-mode squeezed vacuum states can be obtained by mixing a position-squeezed state with a momentum-squeezed state in a 50/50 beam splitter, that is,

$$|r\rangle_{2\text{sq}} = \hat{B}_{12}\left(-\frac{\pi}{4}\right)\left[|-r\rangle_{\text{sq}} \otimes |r\rangle_{\text{sq}}\right]. \tag{4.127}$$

Note that the first beam splitter disappears because the two-mode vacuum state $|0\rangle \otimes |0\rangle$ is invariant under passive transformations.

As a second example, note that, for a single mode, the only passive transformations are the rotations in phase space, which means that an arbitrary single-mode Gaussian unitary can be written as the concatenation of a phase shift, a squeezing operation, a second phase shift, and a final displacement, that is,

$$\hat{U}_{\text{G}}(\alpha, \theta, r, \phi) = \hat{D}(\alpha)\hat{R}(\theta)\hat{S}(r)\hat{R}(\phi). \tag{4.128}$$

Now, it is quite intuitive (and we formalize it in the next section) that any Gaussian state $\hat{\rho}_{\text{G}}$ having von Neumann entropy S_0 can be obtained by applying a unitary transformation on the thermal state $\hat{\rho}_{\text{th}}(\bar{n}_0)$ with the same entropy, $S_{\text{th}}(\bar{n}_0) = S_0$, that is,

$$\hat{\rho}_{\text{G}}(\alpha, \theta, r, \bar{n}_0) = \hat{D}(\alpha)\hat{R}(\theta)\hat{S}(r)\hat{\rho}_{\text{th}}(\bar{n}_0)\hat{S}^{\dagger}(r)\hat{R}^{\dagger}(\theta)\hat{D}^{\dagger}(\alpha), \tag{4.129}$$

---

[6] In optics, an interferometer is just a collection of beam splitters which mix optical beams entering through their input ports. They correspond to the most general passive Gaussian unitary, and are described by a concatenation of single-mode phase shifts and two-mode beam splitters.

which implies that the covariance matrix of any single-mode Gaussian state can always be decomposed as

$$V(\theta, r, \bar{n}_0) = (2\bar{n}_0 + 1)\mathcal{R}(\theta)\mathcal{Q}(2r)\mathcal{R}^T(\theta)$$

$$= (2\bar{n}_0 + 1)\begin{bmatrix} \cosh 2r + \cos 2\theta \sinh 2r & -\sin 2\theta \sinh 2r \\ -\sin 2\theta \sinh 2r & \cosh 2r - \cos 2\theta \sinh 2r \end{bmatrix}. \quad (4.130)$$

Note that the first phase shift has disappeared because thermal states are invariant under such transformations.

Williamson's theorem: symplectic eigenvalues and thermal decomposition
A second interesting result of symplectic analysis is *Williamson's theorem* [7], which states that any positive $2N \times 2N$ symmetric matrix V can be brought to its diagonal form $V^\oplus$ by a symplectic transformation $\mathcal{W}$, that is,

$$V = \mathcal{W}V^\oplus\mathcal{W}^T, \quad \text{with} \quad V^\oplus = \bigoplus_{j=1}^{N} \nu_j \mathcal{I}_{2\times2}. \quad (4.131)$$

This theorem has enormous applicability in the world of Gaussian states. Note that, physically, $V^\oplus$ can be seen as the covariance matrix of N independent modes in a thermal state with mean photon numbers $\{\bar{n}_j = (\nu_j - 1)/2\}_{j=1,2,...,N}$, while the symplectic transformation $\mathcal{W}$ corresponds to a Gaussian unitary transformation. Williamson's theorem is then completely equivalent to stating that any N-mode Gaussian state $\hat{\rho}_G(\bar{r}, V)$ can be obtained as

$$\hat{\rho}_G(\bar{r}, V) = \hat{U}_G(\bar{r}, \mathcal{W})\left\{\bigotimes_{j=1}^{N}\hat{\rho}_{th}\left[(\nu_j - 1)/2\right]\right\}\hat{U}_G^\dagger(\bar{r}, \mathcal{W}). \quad (4.132)$$

The set $\{\nu_j\}_{j=1,2,...,N}$ is called the *symplectic spectrum* of V, and each ν_j is a *symplectic eigenvalue*. It is possible to show that the symplectic spectrum of V can be computed as the absolute values of the eigenvalues of the self-adjoint matrix $i\Omega V$. Note that this expression, together with Euler's decomposition (4.128) of general Gaussian unitaries proves that any single-mode Gaussian state can be written as we did in (4.129).

This decomposition is very important, since it allows us to write many properties of Gaussian states and covariance matrices in an easy manner. For example, the condition (4.67) which ensures that V is the covariance matrix of a physical Gaussian state is simply rewritten as

$$V > 0 \qquad \text{and} \qquad \nu_j \geqslant 1 \; \forall \, j, \quad (4.133)$$

which further shows that any state which saturates the physicality conditions ($\nu_j = 1 \; \forall \, j$) is pure according to (4.132). As a second important example, note that as unitary transformations do not change the von Neumann entropy, the entropy of $\hat{\rho}_G(\bar{r}, V)$ can be directly computed as the sum of the entropies of the corresponding thermal states, that is,

$$S[\hat{\rho}_G(\bar{r}, V)] = \sum_{j=1}^{N}S_{th}\left[(\nu_j - 1)/2\right] = \sum_{j=1}^{N}g(\nu_j), \quad (4.134)$$

where we have defined the function

$$g(x) = \left(\frac{x+1}{2}\right)\log\left(\frac{x+1}{2}\right) - \left(\frac{x-1}{2}\right)\log\left(\frac{x-1}{2}\right), \tag{4.135}$$

which is positive and monotonically increasing for $x \geqslant 1$.

It is particularly simple to evaluate the symplectic eigenvalues in the case of one or two modes. In the single-mode case, the trick is to realize that the determinant of the covariance matrix V is invariant under symplectic transformations, and hence the sole symplectic eigenvalue reads in this case

$$\nu = \sqrt{\det V}. \tag{4.136}$$

For two modes, let us write the covariance matrix in the block form

$$V = \begin{bmatrix} A & C \\ C^T & B \end{bmatrix}, \tag{4.137}$$

where $A = A^T$, $B = B^T$, and C are 2×2 real matrices. In this case, there is an extra symplectic invariant [8] denoted by $\Delta(V) = \det A + \det B + 2 \det C$, and hence the symplectic eigenvalues of a general two-mode Gaussian state can be obtained from

$$\det V = \nu_+^2 \nu_-^2 \qquad \text{and} \qquad \Delta(V) = \nu_+^2 + \nu_-^2, \tag{4.138}$$

leading to

$$\nu_{\pm}^2 = \frac{\Delta(V) \pm \sqrt{\Delta^2(V) - 4 \det V}}{2}. \tag{4.139}$$

In terms of the two-mode symplectic invariants, the second condition in (4.133) is rewritten as

$$\det V \geqslant 1 \qquad \text{and} \qquad \Delta(V) \leqslant 1 + \det V. \tag{4.140}$$

It is particularly relevant the case in which the covariance matrix of the two-mode Gaussian state is in the so-called *standard form*

$$V = \begin{bmatrix} a & 0 & c_1 & 0 \\ 0 & a & 0 & c_2 \\ c_1 & 0 & b & 0 \\ 0 & c_2 & 0 & b \end{bmatrix}. \tag{4.141}$$

Indeed, it is possible to show [9] that the covariance matrix of any bipartite Gaussian state can be brought to this standard form via a local Gaussian unitary transformation $\hat{U}_G = \hat{U}_G(\mathbf{0}, S_1) \otimes \hat{U}_G(\mathbf{0}, S_2)$. The symplectic eigenvalues take a particularly simple form for states which are invariant under correlated or anticorrelated phase shifts of the type $\hat{R}^{(\pm)}(\theta) = \hat{R}(\theta) \otimes \hat{R}(\pm\theta)$, corresponding to states with $c_1 = \pm c_2 = c > 0$, respectively, whose covariance matrices we will denote by

$$V^{(\pm)} = \begin{bmatrix} a & 0 & c & 0 \\ 0 & a & 0 & \pm c \\ c & 0 & b & 0 \\ 0 & \pm c & 0 & b \end{bmatrix}, \tag{4.142}$$

and we will assume $a \geqslant b$ for definiteness and without loss of generality. In the first case, the symplectic eigenvalues read

$$\nu_\pm^{(+)} = \frac{a + b \pm \sqrt{(a-b)^2 + 4c^2}}{2}, \tag{4.143}$$

while in the second case they are

$$\nu_\pm^{(-)} = \frac{\sqrt{(a+b)^2 - 4c^2} \pm (a-b)}{2}. \tag{4.144}$$

Together with the trivial conditions $a \geqslant 1$ and $b \geqslant 1$ obtained from demanding the reduced states for modes 1 and 2 to be physical, the physicality conditions $\nu_-^{(\pm)} \geqslant 1$ imply then $(a \mp 1)(b - 1) \geqslant c^2$ for states invariant under $\hat{R}^{(\pm)}$. Note that $V^{(\pm)}$ have both the same doubly degenerate eigenvalues $\lambda_\pm = (a + b \pm \sqrt{(a-b)^2 + 4c^2})/2$, and hence the condition for their positivity is $ab \geqslant c^2$, which is always granted from the conditions obtained before.

4.4.4 Gaussian bipartite states and Gaussian entanglement

In chapter 2 we introduced the concept of entanglement as correlations between two systems A and B which go beyond the ones allowed classically. In this section we particularize those ideas to continuous-variable states, with emphasis on Gaussian states. In the following we consider only two modes, that is, a 1×1 continuous-variable system, although we shall briefly discuss general $N \times M$ systems as well.

Einstein–Podolsky–Rosen and the birth of entanglement
Even though Einstein is considered as one of the founding fathers of quantum mechanics, he always felt uncomfortable with its probabilistic character. As a result of this criticism, in 1935, with Podolsky and Rosen, he introduced an argument which was supposed to tumble down the foundations of quantum mechanics, showing, in particular, how the theory was both *incomplete* and *inconsistent with causality* [10]. Looking from our current perspective, it is quite ironic how the very same ideas they introduced, far from destroying the theory, are now the ones that fuel many of the most promising applications and deep results in quantum physics.

In this section we will review some of the EPR (for Einstein, Podolsky, and Rosen) arguments in an oversimplified manner, just to obtain a feeling for their ideas and the way they introduced, almost without noticing, the concept of entanglement.

For the sake of argument, let us consider the two-mode squeezed vacuum state (4.113) in the unphysical limit of infinite squeezing, which can be written (up to normalization) as

$$|r \to \infty\rangle_{2sq} = \sum_{n=0}^{\infty} |n, n\rangle \equiv |EPR\rangle, \tag{4.145}$$

where we denote the state by $|EPR\rangle$ because it coincides with the state that EPR used in their seminal paper, as we will now discuss. Let us write the state in terms of the momentum and position eigenstates as follows. First, let us simply apply the identity operator $\int_{\mathbb{R}^2} dx_1 dx_2 |x_1, x_2\rangle\langle x_1, x_2|$, where $|x_1, x_2\rangle = |x_1\rangle \otimes |x_2\rangle$, to the state, obtaining

$$|EPR\rangle = \int_{\mathbb{R}^2} dx_1 dx_2 \left(\sum_{n=0}^{\infty} \langle x_1|n\rangle\langle x_2|n\rangle \right) |x_1, x_2\rangle. \tag{4.146}$$

Now, using the reality of the Fock wave functions (4.30) to write $\langle n|x\rangle^* = \langle n|x\rangle$, the completeness relation of the Fock basis, and the Dirac normalization of the position eigenstates, we turn the former expression into

$$|EPR\rangle = \int_{\mathbb{R}} dx |x, x\rangle = \int_{\mathbb{R}} dp |p, -p\rangle, \tag{4.147}$$

where the last equality is straightforwardly proved by using the Fourier transform relation between the position and momentum eigenbasis. It can be checked that this is precisely the state introduced by EPR in [10].

EPR argue then as follows. Modes 1 and 2 are given, respectively, to *Alice* and *Bob*, two observers placed at distant locations, so that they are not able to interact. Imagine that Alice measures the position and obtains the result[7] x_0; according to quantum mechanics the state of the oscillators collapses to $|x_0, x_0\rangle$, and hence, any subsequent measurement of the position performed by Bob will reveal that his mode has a well defined position x_0. However, Alice could have measured the momentum instead, obtaining for example the result p_0; in this case, quantum mechanics says that the state would have collapsed to $|p_0, -p_0\rangle$, after which Bob would have concluded that his mode had a definite momentum $-p_0$. Now, and this is the center of the argument, *assuming that nothing Alice may do can alter the physical state of Bob's mode* (the modes are separated, even *space-like* or *causally* during the life of Alice and Bob if we like!), one must conclude that Bob's mode must have had well defined values of both its position and momentum from the beginning, hence *violating the quantum-mechanical uncertainty relation* $\Delta X_2 \Delta P_2 \geqslant 1$, and showing that quantum mechanics is inconsistent.

Even though it seems a completely reasonable statement (in particular in 1935, just a decade after the true birth of quantum mechanics), the center of their argument can also be seen as its flaw. The reason is that the state of the system is

---

[7] Again, this is an idealized situation used just for the sake of argument, since having a continuous spectrum, a measurement of $\hat{X}$ cannot give a definite number x_0 as discussed in section 1.3.7.

not an *element of reality* (in EPR's words), it is just a mathematically convenient object which describes the statistics that would be obtained if a physical observable were measured. Consequently, causality does not apply to it: *the actions of Alice can indeed alter Bob's state*, even if these are causally disconnected. Of course, a completely different matter is whether Alice and Bob can use this *spooky action at a distance* (in Einstein's words) to transmit information superluminally. Even though there is no rigorous proof for the negative answer to this question, such a violation of causality has never been observed or even predicted beyond doubt, and hence, most physicists believe that despite non-local effects at the level of states, quantum mechanics cannot violate causality in any way.

The EPR work is the very best example of how one can make advances in a theory by trying to disprove it, since in order to do so one needs to understand it at the deepest level. Even if their motivation was based on 'incorrect' prejudices, they were the first ones to realize that in quantum mechanics it is possible to create correlations which go beyond those admitted in the classical world. States with such property were coined *entangled states* by Schrödinger, who paradoxically was also supportive of the EPR ideas, and a strong believer of the incompleteness of quantum mechanics. Even though there is still a great deal of valuable effort directed towards developing a coherent interpretation of quantum mechanics, over the last few decades physicists have stopped looking at entangled states as the puzzle EPR suggested they were, and started searching for possible applications of them to various problems. Bell was the first who realized the potential of such states in the 1960s, proving that they could be used to rule out the incompleteness of quantum mechanics [11, 12] (exactly the opposite of what EPR created them for!), or, in other words, to prove that the probabilistic character of quantum mechanics does not come from some missing information we fail to account for (at least not without abandoning causality), but from a probabilistic character of nature itself[8]. Throughout the last decades, entangled states have been shown to be a resource for remarkable applications that would be impossible to develop was the world ruled simply by classical physics.

Let us now move to the characterization of entangled states in the continuous-variable regime.

Entanglement criteria for continuous-variable systems
As explained in chapter 2, the Peres–Horodecki criterion, that is, the positivity of the partial transpose of the state, is a necessary condition for a state to be separable. It turns out that it is also a sufficient criterion for 1×1 Gaussian states, and we proceed now to explain how to evaluate it for this special class of states.

It is straightforward to prove from (4.32) that for continuous variables, transposition is equivalent to changing the sign of the momenta [19]. Consequently, the partial transposition operation corresponds to a change of sign in the momentum of

---

[8] Strictly speaking, he proved that no *local hidden-variable theory* is consistent with the predictions of quantum mechanics. These predictions started being tested soon after Bell's discovery [13–15], but we had to wait until very recently [16–18] to see experiments ruling in favor of quantum mechanics beyond any reasonable doubt.

the corresponding modes. In the case of a 1×1 Gaussian state $\hat{\rho}_G(\bar{r}, V)$, this means that partial transposition of the second mode turns the state into $\hat{\rho}_G(\tilde{r}, \tilde{V})$ with

$$\tilde{r} = (I_{2\times 2} \oplus Z)\bar{r} \qquad \text{and} \qquad \tilde{V} = (I_{2\times 2} \oplus Z)V(I_{2\times 2} \oplus Z). \qquad (4.148)$$

The Peres–Horodecki criterion is then reduced to checking whether $\tilde{V}$ is a physical covariance matrix. It is not difficult to prove that $\tilde{V}$ is positive definite and, hence, the criterion is equivalent to checking the condition $\tilde{V} + i\Omega \geqslant 0$, or, equivalently, $\tilde{\nu}_+ \geqslant 1$ in terms of the symplectic eigenvalues of $\tilde{V}$ (note that being symmetric and real, $\tilde{V}$ satisfies Williamson's theorem as well).

As a first interesting example, we consider the states invariant under correlated or anticorrelated phase shifts $\hat{R}^{(\pm)}$, whose covariance matrices can be written in the standard form (4.142). In this case, it is simple to check that partial transposition simply maps one covariance matrix to the other, that is, $\tilde{V}^{(\pm)} = V^{(\mp)}$. Hence, the separability conditions for $V^{(\pm)}$, become the physicality conditions for $V^{(\mp)}$; in other words, $V^{(\pm)}$ corresponds to a separable state if and only if $(a \pm 1)(b - 1) \geqslant c^2$. This implies that states of the type $V^{(+)}$ are always separable, since they must satisfy the physicality condition $c^2 \leqslant (a - 1)(b - 1)$, meaning that $(a + 1)(b - 1)$ cannot ever be smaller than c^2.

Let us consider now a more concrete example consisting of a two-mode squeezed thermal state $\hat{\rho}_{th2sq} = \hat{S}_{12}(r)[\hat{\rho}_{th}(\bar{n}) \otimes \hat{\rho}_{th}(\bar{n})]\hat{S}_{12}^{\dagger}(r)$, whose covariance matrix is

$$V_{th2sq}(r, \bar{n}) = Q_{12}(r)[V_{th}(\bar{n}) \oplus V_{th}(\bar{n})]Q_{12}^{T}(r) = (2\bar{n} + 1)V_{2sq}(r). \qquad (4.149)$$

In this case we obtain

$$\tilde{V}_{th2sq} = (2\bar{n} + 1)\begin{bmatrix} I_{2\times 2}\cosh 2r & I_{2\times 2}\sinh 2r \\ I_{2\times 2}\sinh 2r & I_{2\times 2}\cosh 2r \end{bmatrix}, \qquad (4.150)$$

which has

$$\det\{\tilde{V}_{th2sq}\} = (2\bar{n} + 1)^4 \qquad \text{and} \qquad \Delta(\tilde{V}_{th2sq}) = 2(2\bar{n} + 1)^2\cosh(4r), \qquad (4.151)$$

and therefore symplectic eigenvalues (4.139)

$$\tilde{\nu}_{\pm} = (2\bar{n} + 1)^2 e^{\pm 4r}. \qquad (4.152)$$

For 'zero temperature', $\bar{n} = 0$, we see that $\tilde{\nu}_- < 1$ for any $r > 0$, showing that the two-mode squeezed vacuum state $|r\rangle_{2sq}$ is an entangled state. But for a given squeezing parameter r, we see that there exists a critical thermal occupation $\bar{n}_{crit} = [\exp(2r) - 1]/2$, above which thermal noise is able to destroy the quantum correlations, as $\tilde{\nu}_-$ becomes larger than 1.

From an experimental point of view, the Peres–Horodecki criterion is quite demanding, as it requires the full reconstruction of the state, or at least the full covariance matrix in the case of Gaussian states. However, there is a simpler separability criterion which requires only the analysis of the variance of a suitable pair of joint quadratures (which can be checked experimentally via two homodyne measurements, as we will see later). According to this criterion, which was introduced simultaneously by Duan, Giedke, Cirac, and Zoller [9] and by Simon [19], a sufficient condition for a 1×1 continuous-variable state to be entangled is

$$W^{\phi_1,\phi_2}(\kappa) = V\left(\frac{\kappa \hat{X}_1^{\phi_1} - \kappa^{-1}\hat{X}_2^{\phi_2}}{\sqrt{2}}\right) + V\left(\frac{\kappa \hat{P}_1^{\phi_1} + \kappa^{-1}\hat{P}_2^{\phi_2}}{\sqrt{2}}\right) < \kappa^2 + \kappa^{-2}, \quad (4.153)$$

for some combination of the real parameters ϕ_1, ϕ_2, and $\kappa \neq 0$. In this expression we have defined

$$\hat{X}_j^{\varphi} = \hat{R}_j^{\dagger}(\varphi)\hat{X}_j\hat{R}_j(\varphi) = e^{i\varphi}\hat{a}_j^{\dagger} + e^{-i\varphi}\hat{a}_j = \hat{X}_j \cos\varphi + \hat{P}_j \sin\varphi, \quad (4.154a)$$

$$\hat{P}_j^{\varphi} = \hat{R}_j^{\dagger}(\varphi)\hat{P}_j\hat{R}_j(\varphi) = i\left(e^{i\varphi}\hat{a}_j^{\dagger} - e^{-i\varphi}\hat{a}_j\right) = \hat{P}_j \cos\varphi - \hat{X}_j \sin\varphi, \quad (4.154b)$$

where $\hat{R}_j(\varphi) = \exp(-i\varphi \hat{a}_j^{\dagger}\hat{a}_j)$ induces a phase shift in the corresponding mode, so that the operators above are just the position and momentum quadratures in a rotated phase-space frame. The criterion becomes also a necessary condition for Gaussian states.

Note that for covariance matrices written in the standard form (4.141), the so-called *witness* $W^{\phi_1,\phi_2}(\kappa)$ reduces to

$$W^{\phi_1,\phi_2}(\kappa) = \kappa^2 a + \kappa^{-2} b + (c_2 - c_1)\cos(\phi_1 + \phi_2), \quad (4.155)$$

again showing that states of the type $c_1 = c_2$, that is, states invariant under correlated phase shifts, cannot be entangled, since a and b are larger than 1.

Let us consider the example of the two-mode squeezed vacuum state $|r\rangle_{2sq}$. As its covariance matrix (4.114) is already in standard form, we obtain

$$W^{\phi_1,\phi_2}(\kappa) = (\kappa^2 + \kappa^{-2}) \cosh 2r - 2\cos(\phi_1 + \phi_2)\sinh 2r. \quad (4.156)$$

We see that for $\phi_1 = -\phi_2 \equiv \phi$ and $\kappa = 1$ the witness reads

$$W^{\phi,-\phi}(1) = 2\exp(-2r), \quad (4.157)$$

which is clearly below 2 for every $r > 0$, hence showing once again that $|r\rangle_{2sq}$ is indeed an entangled state. Note, in particular, that for $\phi = 0$ the witness is a sum of the variances of the combined quadratures $\hat{X}_- = (\hat{X}_1 - \hat{X}_2)/\sqrt{2}$ and $\hat{P}_+ = (\hat{P}_1 + \hat{P}_2)/\sqrt{2}$. When the modes are in a vacuum or a coherent state, we have $V(X_-) = V(P_+) = 1$, while in the two-mode squeezed vacuum state the variances are squeezed below this coherent level, $V(X_-) = V(P_+) = \exp(-2r)$. Hence, this state is characterized by the presence of quantum correlations between the modes' positions and anticorrelations between their momenta, as clearly appreciated from its form (4.147) in the $r \to \infty$ limit, and from figure 4.4.

Let us finally stress that a necessary and sufficient criterion for separability has been found for the Gaussian states of a general $N \times M$ bipartite continuous-variable system. This criterion [20] states that the Gaussian state $\hat{\rho}_G(\bar{\mathbf{r}}, V)$ is separable if and

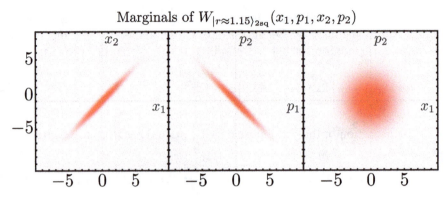

Figure 4.4. Density plot of some characteristic marginals of the Wigner function corresponding to a two-mode squeezed state $|r \approx 1.15\rangle_{2sq}$. In particular, we show $\int_{\mathbb{R}^2} dp_1 dp_2\, W_{|r\rangle_{2sq}}(\mathbf{r})$, $\int_{\mathbb{R}^2} dx_1 dx_2\, W_{|r\rangle_{2sq}}(\mathbf{r})$, and $\int_{\mathbb{R}^2} dp_1 dx_2\, W_{|r\rangle_{2sq}}(\mathbf{r})$, in the left, central, and right panels, respectively. The position correlations and momentum anticorrelations between the modes are apparent, while no special correlation is found between the position of one and the momentum of the other.

only if there exists a pair of matrices V_A and V_B with dimensions $2N \times 2N$ and $2M \times 2M$, respectively, for which

$$V \geqslant V_A \oplus V_B. \tag{4.158}$$

Of course, this criterion is quite difficult to handle in practice, but fortunately an equivalent, operationally friendly criterion was introduced by Giedke, Kraus, Lewenstein, and Cirac, based on the concept of *non-linear maps*. However, we will not go through this criterion which can be consulted in [21], or in the original reference [22].

Quantification of continuous-variable entanglement
Let us move now to the quantification of the entanglement present in a continuous-variable state. As we commented in chapter 2, this problem has been fully solved only for pure states, for which the entanglement entropy is the unique measure of quantum correlations. For mixed states, however, we have not found a completely satisfactory measure (the distillable entanglement and the entanglement of formation are reasonable measures, but cannot be computed for most states, while the logarithmic negativity is easy to compute but does not satisfy all the conditions required of a proper entanglement measure), not even for the reduced class of continuous-variable Gaussian states.

In the following we explain how to compute the entanglement entropy and the logarithmic negativity for 1×1 Gaussian states $\hat{\rho}_G(\bar{\mathbf{r}}, V)$, whose covariance matrix may be written in the same block form as before

$$V = \begin{bmatrix} A & C \\ C^T & B \end{bmatrix}, \tag{4.159}$$

where $A = A^T$, $B = B^T$, and C are 2×2 real matrices (the generalization to $N \times M$ Gaussian states is straightforward).

As commented in (4.69), tracing out one mode of a Gaussian state is equivalent to removing the corresponding rows and columns from the covariance matrix. Hence, in order to evaluate the entanglement entropy of the Gaussian state $\hat{\rho}_G(\bar{\mathbf{r}}, V)$ having covariance matrix (4.159), one just needs to evaluate the entropy of the single-mode covariance matrix A. This matrix has $\nu = \sqrt{\det A}$ as its sole symplectic eigenvalue, and therefore, based on (4.134), its entropy, and hence the entanglement entropy of the corresponding two-mode Gaussian state, is given by

$$E\big[\hat{\rho}_G(\bar{\mathbf{r}}, V)\big] = g\big(\sqrt{\det A}\,\big), \tag{4.160}$$

where the function $g(x)$ was defined in (4.135).

As a pure bipartite state, we can quantify the entanglement of the two-mode squeezed vacuum state $|r\rangle_{2\mathrm{sq}}$ via its entanglement entropy, which in this case reads

$$E\big[|r\rangle_{2\mathrm{sq}}\big] = g(\cosh 2r), \tag{4.161}$$

and is nothing but the entropy of the reduced thermal state. Starting at zero for $r = 0$, this is a monotonically increasing function of r, as expected.

For mixed states the entanglement entropy is not even an entanglement monotone, and hence, it cannot be considered a proper entanglement measure for such states. One then has to consider other measures, and here we focus on the logarithmic negativity $E_N[\hat{\rho}]$. It is possible to show [23] that for an arbitrary Gaussian state $\hat{\rho}_G(\bar{\mathbf{r}}, V)$, this entanglement measure can be computed as

$$E_N\big[\hat{\rho}_G(\bar{\mathbf{r}}, V)\big] = \sum_j F(\tilde{\nu}_j), \tag{4.162}$$

where

$$F(x) = \begin{cases} -\log x & x < 1 \\ 0 & x \geqslant 1 \end{cases}, \tag{4.163}$$

and $\{\tilde{\nu}_j\}_j$ is the symplectic spectrum of the covariance matrix $\tilde{V}$ corresponding to the partial transposition of $\hat{\rho}_G(\bar{\mathbf{r}}, V)$, which is defined in (4.148) for a 1×1 system.

In the case of the two-mode squeezed thermal state (4.149), we obtain $E_N[\hat{\rho}_{\mathrm{th2sq}}] = 2r - \log_e(2\bar{n} + 1)$ below the critical thermal occupation number $(\bar{n} < \bar{n}_{\mathrm{crit}})$.

4.4.5 Gaussian channels

General definition

In section 3.1.1 we introduced general *quantum channels* as trace-preserving quantum operations. Applied to continuous-variable systems, we say that such operations are *Gaussian channels* when they map Gaussian states into Gaussian states. It is in this context where the name 'channel' is best suited, since Gaussian channels indeed model some of the most important *communication channels* such as fibers, wires, and amplifiers.

As we saw, a way of characterizing an arbitrary trace-preserving quantum operation $\mathcal{E}$ is by giving a complete set of Kraus operators $\{\hat{E}_k\}_{k=1,2,\ldots,K}$ which transform a state $\hat{\rho}$ into the state

$$\hat{\rho} \to \mathcal{E}[\hat{\rho}] = \sum_{k=1}^{K} \hat{E}_k \hat{\rho} \hat{E}_k^{\dagger}. \tag{4.164}$$

Gaussian channels, on the other hand, can be characterized by their action on the first and second moments of the state. In particular, similarly to Gaussian unitaries, and as will be clear from the following discussion, Gaussian channels acting on N modes are characterized by a vector $\mathbf{d} \in \mathbb{R}^{2N}$, plus two real $2N \times 2N$ matrices $\mathcal{K}$ and $\mathcal{N}$, which transform the mean vector $\bar{\mathbf{r}}$ and covariance matrix V of the state as

$$\bar{\mathbf{r}} \to \mathcal{K}\bar{\mathbf{r}} + \mathbf{d} \qquad \text{and} \qquad V \to \mathcal{K}V\mathcal{K}^{T} + \mathcal{N}. \tag{4.165}$$

The matrices $\mathcal{K}$ and $\mathcal{N}$ must satisfy certain conditions in order to correspond to a true trace-preserving quantum operation. First, as the covariance matrix is symmetric, so must be the matrix $\mathcal{N}$. Second, in order to be a completely-positive map, they have to satisfy the following restriction [24]

$$\mathcal{N} + i\Omega - i\mathcal{K}\Omega\mathcal{K}^{T} \geqslant 0. \tag{4.166}$$

Note that Gaussian unitaries correspond to a Gaussian channel for which $\mathcal{N}$ is zero, and $\mathcal{K}$ is symplectic. It is possible to show that for single-mode channels ($N = 1$) this last condition can be rewritten as

$$\mathcal{N} \geqslant 0 \qquad \text{and} \qquad \det \mathcal{N} \geqslant (\det \mathcal{K} - 1)^2. \tag{4.167}$$

Roughly speaking, $\mathcal{K}$ plays the role of the amplification or attenuation of the channel (plus a possible rotation), while $\mathcal{N}$ includes any source of quantum or classical noise added by the channel. We shall return to their physical interpretation in the next section via a particular example.

Indeed, it is quite simple to understand why Gaussian channels correspond to a transformation of the type (4.165). To this aim, we just need to remember that any trace-preserving quantum operation can be seen as the reduced dynamics induced by a unitary transformation acting on the system and some environment in a pure state (Stinespring dilation). It is obvious that in order for the channel to be Gaussian, both the state of the environment $|\psi_E\rangle$ and the joint unitary transformation $\hat{U}_{\mathrm{G}}(\mathbf{s}, S)$ must be Gaussian. Moreover, as every pure Gaussian state is connected to the vacuum state via some Gaussian unitary transformation which can be included in the joint unitary $\hat{U}_{\mathrm{G}}(\mathbf{s}, S)$, we can take the initial state of the environment as the multi-mode vacuum state, that is,

$$|\psi_E\rangle = \bigotimes_{j=1}^{N_E} |0\rangle \equiv |\mathrm{vac}\rangle, \tag{4.168}$$

where we have assumed that the environment consists in N_E modes. The state $\hat{\rho}$ of the system is then transformed into

$$\hat{\rho}' = \mathrm{tr}_E\left\{\hat{U}_{\mathrm{G}}(\mathbf{s}, S)[\hat{\rho} \otimes |\mathrm{vac}\rangle\langle\mathrm{vac}|]\hat{U}_{\mathrm{G}}^{\dagger}(\mathbf{s}, S)\right\}. \tag{4.169}$$

Let us write the Gaussian parameters associated to the joint unitary as

$$\mathbf{s} = \mathrm{col}(\mathbf{d}, \mathbf{d}_E) \qquad \text{and} \qquad S = \begin{bmatrix} S_S & S_{SE} \\ S_{ES} & S_E \end{bmatrix}, \qquad (4.170)$$

where $\mathbf{d} \in \mathbb{R}^{2N}$, and the real matrices S_S, S_E, S_{SE}, and S_{ES}, have dimensions $2N \times 2N$, $2N_E \times 2N_E$, $2N \times 2N_E$, and $2N_E \times 2N$, respectively. Let us write also the mean and the covariance matrix of the initial separable joint state of the system plus the environment as

$$\bar{\mathbf{r}}_{SE} = \mathrm{col}(\bar{\mathbf{r}}, \mathbf{0}) \qquad \text{and} \qquad V_{SE} = V \oplus \mathcal{I}_{2N_E \times 2N_E}. \qquad (4.171)$$

After the unitary, these are transformed into

$$\bar{\mathbf{r}}'_{SE} = S\,\bar{\mathbf{r}}_{SE} = \mathrm{col}(S_S\bar{\mathbf{r}} + \mathbf{d}, S_{ES}\bar{\mathbf{r}} + \mathbf{d}_E), \qquad (4.172a)$$

$$V'_{SE} = S V_{SE} S^T = \begin{bmatrix} S_S V S_S^T + S_{SE} S_{SE}^T & S_S V S_S + S_{SE} S_E^T \\ S_{ES} V S_S^T + S_E S_{SE}^T & S_{ES} V S_{ES}^T + S_E S_E^T \end{bmatrix}, \qquad (4.172b)$$

so that by tracing out the environment, the transformation onto the mean vector and the covariance matrix of the system is

$$\bar{\mathbf{r}}' = S_S\bar{\mathbf{r}} + \mathbf{d} \qquad \text{and} \qquad V' = S_S V S_S^T + S_{SE} S_{SE}^T, \qquad (4.173)$$

which is exactly the type of transformation introduced in (4.165), where we now make the identifications

$$\mathcal{K} = S_S \qquad \text{and} \qquad \mathcal{N} = S_{SE} S_{SE}^T. \qquad (4.174)$$

Interestingly, it can be proved that the Stinespring dilation of any Gaussian channel can be generated by choosing an environment with less than twice the number of modes of the system, that is, $N_E \leqslant 2N$ [25, 26].

Note that the Gaussianity is a property of the channel, not of the state of the system, that is, one can consider the action of the Gaussian channel on non-Gaussian states. Indeed, the transformation of a general state $\hat{\rho}$ after passing through the channel receives a very simple description in terms of characteristic functions. To see this, using the *block inversion formula* [27], let us write the inverse of the symplectic matrix S as

$$S^{-1} = \begin{bmatrix} \mathcal{T}_S & \mathcal{T}_{SE} \\ \mathcal{T}_{ES} & \mathcal{T}_E \end{bmatrix} = \begin{bmatrix} (S/S_E)^{-1} & -S_S^{-1}S_{SE}(S/S_E)^{-1} \\ -(S/S_E)^{-1}S_{ES}S_S^{-1} & (S/S_S)^{-1} \end{bmatrix}, \qquad (4.175)$$

where

$$S/S_S = S_E - S_{ES}S_S^{-1}S_{SE} \qquad \text{and} \qquad S/S_E = S_S - S_{SE}S_E^{-1}S_{ES}, \qquad (4.176)$$

are the so-called *Schur complements* of S_S and S_E, respectively. Let us also denote by **r** and $\mathbf{r}_E$ the phase-space coordinates of the system and environment modes, respectively, so that the initial characteristic function can be written as

$$\chi_0(\mathbf{r}, \mathbf{r}_E) = \chi_{\hat{\rho}}(\mathbf{r})\chi_{\text{vac}}(\mathbf{r}_E). \tag{4.177}$$

Then, recalling the transformation of the characteristic function under Gaussian unitaries (4.86), and after tracing out the environmental modes (that is, setting to zero their phase-space variables), we obtain the output characteristic function

$$\chi_{\mathcal{E}[\hat{\rho}]}(\mathbf{r}) = e^{-\frac{i}{2}\mathbf{r}^T\Omega\mathbf{d}}\chi_{\hat{\rho}}(\mathcal{T}_S\mathbf{r})\chi_{\text{vac}}(\mathcal{T}_{ES}\mathbf{r}). \tag{4.178}$$

In the following, we will denote by $\mathcal{E}(\mathcal{K}, \mathcal{N})$ any Gaussian channel, obviating the displacement **d** which, at least for Gaussian states, can actually be generated after the channel via a unitary displacement transformation (4.92), since this does not change the covariance matrix in any way.

An example: phase-insensitive Gaussian channels
There is a particularly simple class of single-mode Gaussian channels which plays an important role in communication technologies: *phase-insensitive Gaussian channels*, which are defined by

$$\mathcal{K} = \sqrt{\tau}\,\mathcal{I}_{2\times2} \qquad \text{and} \qquad \mathcal{N} = \mu\mathcal{I}_{2\times2}, \tag{4.179}$$

where $\tau \geqslant 0$ and $\mu \geqslant 0$ satisfy

$$\mu \geqslant |\tau - 1|, \tag{4.180}$$

by virtue of the positivity condition (4.167). Note that these channels are called 'phase insensitive' because they are invariant under rotations in phase space. We will denote them by $C(\tau, \mu)$.

After crossing the channel, the covariance matrix V of any state is transformed into

$$V' = \mathcal{K}V\mathcal{K}^T + \mathcal{N} = \begin{bmatrix} \tau V_{11} + \mu & \tau V_{12} \\ \tau V_{21} & \tau V_{22} + \mu \end{bmatrix}, \tag{4.181}$$

and hence

$$\text{tr}\{V'\} = \tau\,\text{tr}\{V\} + 2\mu, \tag{4.182}$$

$$\det\{V'\} = \tau^2\det\{V\} + \mu(\tau\,\text{tr}\{V\} + \mu). \tag{4.183}$$

Taking into account that the mean photon number is proportional to the trace of the covariance matrix, see (4.66), while the von Neumann entropy is a monotonically increasing function of its determinant, see (4.134) and (4.136), we conclude that τ acts as an attenuation (for $0 \leqslant \tau < 1$) or amplification (for $\tau > 1$) factor, while μ adds noise (mixedness) to the state. Note that this implies that quantum mechanics does not allow for the attenuation or amplification of a signal without introducing noise

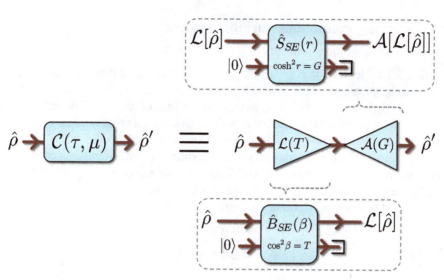

Figure 4.5. Decomposition of a general phase-insensitive Gaussian channel as a concatenation of a pure-loss channel and a quantum-limited amplifier, and Stinespring dilations of the latter.

(at least deterministically, that is, via trace-preserving operations), which comes from the fact that the uncertainty principle must be satisfied at all times.

There are two interesting limiting cases of this class of channels:

- For $0 \leqslant \tau < 1$ and $\mu = 1 - \tau$ we obtain the so-called *pure-loss channels*, which are a good approximation of the fibers used in current optical communication technologies. It is fairly simple to check that the simplest Stinespring dilation of such channels consists in mixing the input mode with a single environmental mode in a beam splitter (4.115a) with mixing angle $\cos^2 \beta = \tau$ (see figure 4.5). In this context, it is usual to call τ the *attenuation factor* or *transmissivity*, T, and the channel is usually denoted by $\mathcal{L}(T)$.

- For $\tau > 1$ and $\mu = \tau - 1$, we obtain the so-called *quantum-limited amplifiers*, which are the less noisy (deterministic) amplifiers that quantum mechanics allows. Again, it is simple to check that the simplest Stinespring dilation of these channels consists in mixing the input mode with a single environmental mode in a two-mode squeezer (4.109) with squeezing parameter satisfying $\cosh^2 r = \tau$ (see figure 4.5). The parameter τ receives here the name *amplification factor* or *gain*, G, and the channel is usually denoted by $\mathcal{A}(G)$.

Any phase-insensitive Gaussian channel can be written as the concatenation of a pure-loss channel and a quantum-limited amplifier, that is, $\mathcal{C}(\tau, \mu) = \mathcal{L}(T) \circ \mathcal{A}(G)$, as depicted in figure 4.5, where

$$\tau = TG \qquad \text{and} \qquad \mu = G(1 - T) + (G - 1). \qquad (4.184)$$

This is trivially proven by directly applying both channel transformations subsequently onto a general covariance matrix, and then comparing the result with (4.181).

4.5 Measuring continuous-variable systems

4.5.1 Description of measurements in phase space

In section 3.2 we learned that the most general measurement that one can perform in a quantum system can always be described by a complete set of trace-decreasing operations $\{\mathcal{E}_j\}_{j=1,2,\ldots,J>1}$, each corresponding to one of the possible measurement outcomes. In this context we defined POVM-based measurements as those generalized measurements whose quantum operations can each be described by a single Kraus operator, which were shown to be the simplest generalizations of projective measurements. In this section we will learn a convenient way of describing this type of generalized measurement for continuous-variable systems.

Let us start with some useful definitions. Given the initial state of a system $\hat{\rho}$, and the POVM $\{\hat{\Pi}_j = \hat{E}_j^\dagger \hat{E}_j\}_{j=1,2,\ldots,J}$, we will denote by

$$\tilde{\rho}_j = \mathcal{E}_j[\hat{\rho}] = \hat{E}_j \hat{\rho} \hat{E}_j^\dagger, \tag{4.185}$$

the unnormalized state obtained after the outcome j appears. Such an outcome appears with probability $p_j = \text{tr}\{\tilde{\rho}_j\}$, and the normalized state of the system reads $\hat{\rho}_j = p_j^{-1}\tilde{\rho}_j$. Similarly, and assuming that the system is described as a collection of N modes, we define the corresponding unnormalized characteristic and Wigner functions as

$$\tilde{\chi}_j(\mathbf{r}) = \text{tr}\left\{\hat{D}(\mathbf{r})\tilde{\rho}_j\right\} \quad \text{and} \quad \tilde{W}_j(\mathbf{r}) = \int_{\mathbb{R}^{2N}} \frac{d^{2N}\mathbf{s}}{(4\pi)^{2N}}\tilde{\chi}_j(\mathbf{s})e^{\frac{i}{2}\mathbf{s}^T\Omega\mathbf{r}}, \tag{4.186}$$

from which the probability of the corresponding outcome can be obtained as

$$p_j = \tilde{\chi}_j(\mathbf{0}) = \int_{\mathbb{R}^{2N}} d^{2N}\mathbf{r}\,\tilde{W}_j(\mathbf{r}), \tag{4.187}$$

and the normalized functions as

$$\chi_j(\mathbf{r}) = \text{tr}\left\{\hat{D}(\mathbf{r})\hat{\rho}_j\right\} = p_j^{-1}\tilde{\chi}_j(\mathbf{r}), \tag{4.188a}$$

$$W_j(\mathbf{r}) = \int_{\mathbb{R}^{2N}} \frac{d^{2N}\mathbf{s}}{(4\pi)^{2N}}\chi_j(\mathbf{s})e^{\frac{i}{2}\mathbf{s}^T\Omega\mathbf{r}} = p_j^{-1}\tilde{W}_j(\mathbf{r}). \tag{4.188b}$$

There are many situations in which the measurement is not applied to the whole system, but only to one of the modes that conform to it. We talk in those cases about *partial measurements*. Moreover, as we will see in the next sections, the measurement performed on a light beam is usually *destructive*, that is, the mode disappears after the measurement is performed, so that one has to trace it out of the system. Assuming that the system has $N + 1$ modes, and that the measurement is applied on the last mode, this means that the (unnormalized) state of the N modes remaining after the measurement will be[9]

---

[9] Note that the cyclic property applies also to the partial trace when the joint operator acts as the identity on the non-traced subspaces. In particular, for a bipartite Hilbert space $\mathcal{H}_A \otimes \mathcal{H}_B$, it is simple to prove (for example by expanding the expression explicitly in a basis) that $\text{tr}_B\{\hat{\rho}(\hat{I}_A \otimes \hat{B})\} = \text{tr}_B\{(\hat{I}_A \otimes \hat{B})\hat{\rho}\}$, where $\hat{\rho}$ is any operator acting on the joint Hilbert space and $\hat{I}_A$ is the identity operator acting on $\mathcal{H}_A$.

$$\tilde{\rho}_j = \mathrm{tr}_{N+1}\left\{\left(\hat{I}_N \otimes \hat{E}_j\right)\hat{\rho}\left(\hat{I}_N \otimes \hat{E}_j^\dagger\right)\right\} = \mathrm{tr}_{N+1}\left\{\left(\hat{I}_N \otimes \hat{\Pi}_j\right)\hat{\rho}\right\}, \qquad (4.189)$$

where $\hat{\rho}$ is the initial state of the $N+1$ modes. The first thing to note is that, obviously, the first N modes only 'feel' the measurement if they share some correlations with the measured mode. On the other hand, an interesting feature of such partial, destructive measurements is that one only needs the POVM $\{\hat{\Pi}_j\}_{j=1,2,...,J}$ to evaluate the final state of the non-measured modes; this is in contrast to non-destructive measurements, which require knowledge of the measurement operators $\{\hat{E}_j\}_{j=1,2,...,J}$ in order to compute the post-measurement state.

Partial measurements are easily described in terms of characteristic and Wigner functions. In the case of the characteristic function the derivation is simple by using its relations with the density operator (4.54):

$$\tilde{\chi}_j(\mathbf{r}_{\{N\}}) = \mathrm{tr}_{\{N\}}\left\{\hat{D}(\mathbf{r}_{\{N\}})\tilde{\rho}_j\right\} = \mathrm{tr}\left\{\left[\hat{D}(\mathbf{r}_{\{N\}}) \otimes \hat{\Pi}_j\right]\hat{\rho}\right\}$$

$$= \int_{\mathbb{R}^{2(N+1)}} \frac{\mathrm{d}^{2(N+1)}\mathbf{s}}{(4\pi)^{N+1}}\chi_{\hat{\rho}}(\mathbf{s})\underbrace{\mathrm{tr}_{\{N\}}\left\{\hat{D}(\mathbf{r}_{\{N\}} - \mathbf{s}_{\{N\}})\right\}}_{(4\pi)^N\delta^{(2N)}(\mathbf{r}_{\{N\}}-\mathbf{s}_{\{N\}})}\underbrace{\mathrm{tr}_{N+1}\left\{\hat{\Pi}_j\hat{D}^\dagger(\mathbf{s}_{N+1})\right\}}_{\chi_{\hat{\Pi}_j}(-\mathbf{s}_{N+1})}$$

$$= \int_{\mathbb{R}^2} \frac{\mathrm{d}^2\mathbf{r}_{N+1}}{4\pi}\chi_{\hat{\rho}}(\mathbf{r})\chi_{\hat{\Pi}_j}(-\mathbf{r}_{N+1}). \qquad (4.190)$$

Using (4.186) and (4.55), we derive now the transformation rule for the Wigner function:

$$\tilde{W}_j(\mathbf{r}_{\{N\}}) = \int_{\mathbb{R}^{2N}} \frac{\mathrm{d}^{2N}\mathbf{s}_{\{N\}}}{(4\pi)^{2N}} \int_{\mathbb{R}^2} \frac{\mathrm{d}^2\mathbf{s}_{N+1}}{4\pi}\chi_{\hat{\rho}}(\mathbf{s}_{\{N\}}, \mathbf{s}_{N+1})\chi_{\hat{\Pi}_j}(-\mathbf{s}_{N+1})e^{\frac{i}{2}\mathbf{s}_{\{N\}}^T\Omega_N\mathbf{r}_{\{N\}}}$$

$$= \int_{\mathbb{R}^2} \mathrm{d}^2\mathbf{r}_{N+1} \int_{\mathbb{R}^{2(N+1)}} \frac{\mathrm{d}^{2(N+1)}\mathbf{s}}{(4\pi)^{2N+1}}\chi_{\hat{\rho}}(\mathbf{s})W_{\hat{\Pi}_j}(\mathbf{r}_{N+1})e^{\frac{i}{2}\mathbf{s}^T\Omega\mathbf{r}}$$

$$= 4\pi \int_{\mathbb{R}^2} \mathrm{d}^2\mathbf{r}_{N+1}W_{\hat{\rho}}(\mathbf{r})W_{\hat{\Pi}_j}(\mathbf{r}_{N+1}). \qquad (4.191)$$

Hence, both for the characteristic and the Wigner functions, the transformation is obtained by multiplying the initial function of the $N+1$ modes by the function associated with the POVM element $\hat{\Pi}_j$ (with the suitable sign in the argument), and integrating out the measured mode.

Let us now go on to describe some particular types of measurements which find many applications in the world of continuous variables, using light modes as the guiding context.

4.5.2 Photodetection: measuring the photon number

The most fundamental measurement technique for light is photodetection. As we shall see with a couple of examples (homodyne detection and on/off detection), virtually any other scheme used for measuring different properties of light makes use of photodetection.

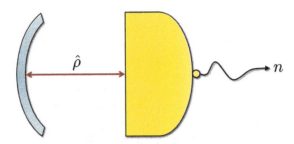

Figure 4.6. Schematic representation of Mollow's idea of photodetection, ideally implementing a measurement of the number operator associated with a single mode.

Ideal photodetection

This technique is based on the photoelectric effect or variations of it[10]. The idea is that when the light beam that we want to detect impinges a metallic surface, it is able to release some of the bound electrons of the metal, which are then collected by an anode. The same happens if light impinges on a semiconductor surface, though in this case instead of becoming free, valence electrons are promoted to the conduction band. The most widely used metallic photodetectors are known as *photo-multiplier tubes*, while those based on semiconducting films are the so-called *avalanche photodiodes*. In both cases, each photon is able to create a single electron, whose associated current would be equally difficult to measure by electronic means. For this reason, each photoelectron is accelerated towards a series of metallic plates at increasing positive voltages, releasing then more electrons via scattering which contribute to generating a measurable electric pulse, the *photopulse*.

It is then customarily said that counting photopulses is equivalent to counting photons, and hence, photodetection is equivalent to a measurement of the number of photons of the light field. This is a highly idealized situation, valid only in some limits which we will try to understand now.

Consider the following model for a perfectly efficient detection scheme. A single-mode field initially in some state $\hat{\rho}$ is kept in continuous interaction with a photodetector during a time interval T. The intuitive picture of such a scenario is shown in figure 4.6: a cavity formed by the photodetector itself and an extra perfectly reflecting mirror contains a single mode. By developing a microscopic model of the detector and its interaction with the light mode, Mollow was able to show that the probability of generating n photoelectrons (equivalently, the probability of observing n photopulses) during the time interval T is given by [28]

$$p_n = \mathrm{tr}\left\{\hat{\rho} : \frac{(1 - \mathrm{e}^{-\kappa T})^n \hat{a}^{\dagger n} \hat{a}^n}{n!} \exp[-(1 - \mathrm{e}^{-\kappa T})\hat{a}^\dagger \hat{a}] : \right\}, \qquad (4.192)$$

---

[10] In the last few years superconducting photodetectors have received a lot of attention. However, we will stick to the photoelectric ones both for simplicity, and because they are the ones currently found in most laboratories around the world.

where κ is some parameter accounting for the light–detector interaction, and the double-dots refer to the operation of normal ordering[11]. Using the operator identity $:\exp[-(1 - e^{-\lambda})\hat{a}^\dagger\hat{a}]: = \exp(-\lambda\hat{a}^\dagger\hat{a})$ [29], and the help of the number state basis $\{|n\rangle\}_{n\in\mathbb{N}}$, it is straightforward to obtain

$$p_n = \sum_{m=n}^{\infty}\langle m|\hat{\rho}|m\rangle\frac{m!}{n!(m-n)!}(1 - e^{-\kappa T})^m(e^{-\kappa T})^{m-n}\xrightarrow[T\gg\kappa^{-1}]{}\langle n|\hat{\rho}|n\rangle, \quad (4.193)$$

and hence, for large enough detection times the number of observed pulses follows the statistics of the number of photons. In other words, this ideal photodetection scheme is equivalent to measuring the number operator $\hat{N} = \hat{a}^\dagger\hat{a}$ as already commented, that is, a projective measurement with projectors $\{\hat{P}_n = |n\rangle\langle n|\}_{n=0,1,2,...}$.

Photodetection in the presence of finite efficiency and dark counts
However, in real photodetectors the condition $T \gg \kappa^{-1}$ is hardly met. One usually defines the *quantum efficiency* $\eta = 1 - e^{-\kappa T}$, which in current photodetectors varies from one wavelength to another, and then, according to (4.193) photodetection is equivalent to a generalized measurement with POVM elements

$$\hat{\Pi}_n = \sum_{m=n}^{\infty}\binom{m}{n}\eta^n(1 - \eta)^{m-n}|n\rangle\langle n|, \quad n = 0, 1, 2, ... \quad (4.194)$$

This POVM-based measurement admits a very simple Stinespring dilation, which is quite convenient to gain some intuition about optical measurement schemes: before arriving at a photodetector with unit quantum efficiency, the optical mode is mixed with an ancillary vacuum mode in a beam splitter of transmissivity $\cos^2\beta = \eta$. In order to prove that this scheme leads to the same POVM as the one associated with a detector with finite efficiency, let us compute the probability of observing n photopulses in the detector. Using the identity [30]

$$\hat{B}(\beta) = e^{\beta\hat{a}\hat{a}_E^\dagger - \beta\hat{a}^\dagger\hat{a}_E} = e^{\hat{a}\hat{a}_E^\dagger\tan\beta}(\cos\beta)^{\hat{a}^\dagger\hat{a} - \hat{a}_E^\dagger\hat{a}_E}e^{-\hat{a}^\dagger\hat{a}_E\tan\beta}, \quad (4.195)$$

and taking into account that

$$e^{\zeta\hat{a}\hat{a}_E^\dagger}|k, 0\rangle = \sum_{j=0}^{k}\frac{\zeta^j}{j!}\hat{a}^j\hat{a}_E^{\dagger j}|k, 0\rangle = \sum_{j=0}^{k}\zeta^j\sqrt{\binom{k}{j}}|k - j, j\rangle, \quad (4.196)$$

the state of the system after the beam splitter can be written as

$$\hat{\rho}_{SE} = \hat{B}(\beta)(\hat{\rho} \otimes |0\rangle\langle 0|)\hat{B}^\dagger(\beta)$$

$$= \sum_{l,m=0}^{\infty}\rho_{lm}\cos^{l+m}\beta\sum_{j=0}^{l}\sum_{k=0}^{m}\sqrt{\binom{l}{j}\binom{m}{k}}\tan^{j+k}\beta|l - j, j\rangle\langle m - k, k|. \quad (4.197)$$

The reduced state of the detected mode then reads,

$$\hat{\rho}' = \text{tr}_E\{\hat{\rho}_{SE}\} = \sum_{l,m=0}^{\infty}\rho_{lm}\cos^{l+m}\beta\sum_{k=0}^{\min\{l,m\}}\sqrt{\binom{l}{k}\binom{m}{k}}\tan^{2k}\beta|l - k\rangle\langle m - k|, \quad (4.198)$$

---

[11] Given any multiplication of creation and annihilation operators, this operation amounts to bringing all the creation (annihilation) operators to the left (right) as if they commute. Hence, for example, $: \hat{a}\hat{a}^\dagger := \hat{a}^\dagger\hat{a}$.

and since the detector is considered ideal, the probability of observing n photopulses is equal to

$$p_n = \langle n|\hat{\rho}'|n\rangle = \sum_{l,m=0}^{\infty} \rho_{lm} \cos^{l+m}\beta \sum_{k=0}^{\min\{l,m\}} \sqrt{\binom{l}{k}\binom{m}{k}} \tan^{2k}\beta \underbrace{\delta_{l-k,n}\delta_{m-k,n}}_{\delta_{lm}\delta_{k,l-n}}$$

$$= \sum_{m=n}^{\infty} \rho_{mm}\left(\frac{m}{m-n}\right)\cos^{2m}\beta \tan^{2(m-n)}\beta, \tag{4.199}$$

which coincides with (4.193) once the identification $\cos^2\beta = \eta$ is performed.

Apart from the finite quantum efficiency, which accounts for the missed photons which do not generate photoelectrons in the detector, there is another source of imperfection that can be understood as the dual to the latter: electrons which are pulled out from the detector without interacting with any photon of the detected mode. One refers to the corresponding photopulses as *dark counts*, and they can be modeled within the previous Stinespring dilation in a very simple way: by assuming that the ancilla mode is not in a vacuum but rather in some other state, say $\hat{\rho}_E$, usually taken as a thermal state. In this scenario, the POVM elements become

$$\hat{\Pi}_n = \text{tr}_E\left\{\hat{B}(\beta)\left(\hat{I} \otimes \hat{\rho}_E\right)\hat{B}^{\dagger}(\beta)\left(|n\rangle\langle n| \otimes \hat{I}\right)\right\}. \tag{4.200}$$

In the following all these imperfections will be ignored, so that we will assume that photodetection is equivalent to a measurement of the number of photons of the field impinging the detector. However, it is important to understand how to deal with these experimental limitations before proposing any interesting theoretical protocol, in case one has the need to perform a more realistic analysis.

4.5.3 Homodyne and heterodyne detection: measuring the quadratures and the annihilation operator

Basic principles behind homodyning

Even though the output of the photodetectors can take only integer values (the number of recorded photopulses), they can be arranged to approximately measure the quadratures of light, which we recall are continuous observables. This arrangement is called *homodyne detection*. The basic scheme is shown in figure 4.7. The mode we want to measure is mixed in a beam splitter with another mode, called the *local oscillator*, which is in a coherent state $|\alpha_{\text{LO}}\rangle$. When the beam splitter is 50/50 the homodyne scheme is said to be *balanced*, and the annihilation operators associated with the modes transformed through the beam splitter are given by

$$\hat{a}_{\pm} = \frac{1}{\sqrt{2}}(\hat{a} \pm \hat{a}_{\text{LO}}), \tag{4.201}$$

where $\hat{a}_{\text{LO}}$ is the annihilation operator of the local oscillator mode. These modes are measured with independent photodetectors, and then the corresponding signals are subtracted. Based on the idealized photodetection picture of the previous section,

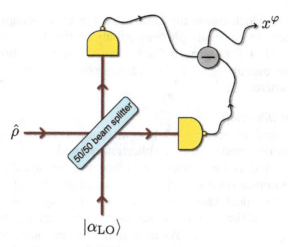

Figure 4.7. Homodyne detection scheme with ideal photodetectors. When the local oscillator is in a strong coherent state, this set up gives access to the quadratures of light.

this scheme is analogous to a measurement of the photon-number difference

$$\hat{N}_D = \hat{a}_+^\dagger \hat{a}_+ - \hat{a}_-^\dagger \hat{a}_- = \hat{a}_{LO}^\dagger \hat{a} + \hat{a}_{LO} \hat{a}^\dagger. \tag{4.202}$$

Taking into account that the local oscillator is in a coherent state with an amplitude $\alpha_{LO} = |\alpha_{LO}| \exp(i\varphi)$, and that it is not correlated with the mode we want to measure, it is not difficult to show that the first moments of this operator can be written as

$$\langle \hat{N}_D \rangle = |\alpha_{LO}| \langle \hat{X}^\varphi \rangle, \tag{4.203a}$$

$$\left\langle \hat{N}_D^2 \right\rangle = |\alpha_{LO}|^2 \left[\left\langle \hat{X}^{\varphi 2} \right\rangle + \frac{\langle \hat{a}^\dagger \hat{a} \rangle}{|\alpha_{LO}|^2} \right], \tag{4.203b}$$

where

$$\hat{X}^\varphi = e^{-i\varphi} \hat{a} + e^{i\varphi} \hat{a}^\dagger, \tag{4.204}$$

is a generalized quadrature which coincides with the position and momentum for $\varphi = 0$ and $\pi/2$, respectively. Hence, in the *strong local oscillator limit* $|\alpha_{LO}|^2 \gg \langle \hat{a}^\dagger \hat{a} \rangle / \langle \hat{X}^{\varphi 2} \rangle$, the output signal of the homodyne scheme has the mean of a quadrature $\hat{X}^\varphi$ of the analyzed mode (the one selected by the phase of the local oscillator), as well as its same variance. Moreover, it is simple but tedious to check that all the moments of $\hat{N}_D$ coincide with those of $\hat{X}^\varphi$ in the strong local oscillator limit, and therefore, balanced homodyne detection can be seen as a measurement of the corresponding quadrature.

It might be difficult to accept that a measurement of $\hat{N}_D$, which has a discrete spectrum, can be equivalent to a measurement of $\hat{X}^\varphi$, which has a continuous spectrum. The reconciliation between these two pictures comes from the strong local oscillator condition, which essentially means that the local oscillator is very intense and, therefore, there are so many photons impinging the detectors that the photo-pulses are generated at a rate much faster than the response time of the photo-detectors, and the output signal is, basically, experienced by the observer as a continuous photocurrent.

Partial homodyne detection

Just as we explained in section 4.5.1, sometimes it is interesting to apply a measurement onto one mode out of a collection of modes (say the last mode of a system with $N + 1$ modes); this is what we defined as a partial measurement. Partial homodyne detection receives a very simple treatment in terms of Wigner functions. For example, in the ideal case explained above (strong local oscillator limit), homodyne detection of the position quadrature is described by the continuous set of projectors $\{\hat{P}(x) = |x\rangle\langle x|\}_{x \in \mathbb{R}}$. Assume that the measurement pops out the outcome[12] x_0, so that, according to (4.191) and (4.190), the unnormalized character-istic and Wigner functions of the remaining N modes collapse to

$$\tilde{\chi}_{x_0}(\mathbf{r}_{\{N\}}) = \int_{\mathbb{R}^2} \frac{d^2 \mathbf{r}_{N+1}}{4\pi} \chi_{\hat{\rho}}(\mathbf{r}) \chi_{|x_0\rangle\langle x_0|}(-\mathbf{r}_{N+1}), \qquad (4.205a)$$

$$\tilde{W}_{x_0}(\mathbf{r}_{\{N\}}) = 4\pi \int_{\mathbb{R}^2} d^2 \mathbf{r}_{N+1} W_{\hat{\rho}}(\mathbf{r}) W_{|x_0\rangle\langle x_0|}(\mathbf{r}_{N+1}), \qquad (4.205b)$$

where $\hat{\rho}$ is the initial state of the $N + 1$ modes. The characteristic function of the projector $|x_0\rangle\langle x_0|$ is easily found as

$$\chi_{|x_0\rangle\langle x_0|}(\mathbf{r}) = \text{tr}\left\{ \hat{D}(\mathbf{r})|x_0\rangle\langle x_0| \right\} = e^{-ipx/4}\langle x_0|e^{ip\hat{X}/2}e^{-ix\hat{P}/2}|x_0\rangle$$

$$= e^{-ipx/4}e^{ipx_0/2} \underbrace{\int_{\mathbb{R}} dp_0 e^{-ixp_0/2}}_{4\pi\delta(x)} \underbrace{\langle x_0|p_0\rangle\langle p_0|x_0\rangle}_{1/4\pi} = \delta(x)e^{ipx_0/2}, \qquad (4.206)$$

where we have used (4.19), while Fourier transforming this expression we obtain the corresponding Wigner function

$$W_{|x_0\rangle\langle x_0|}(x, p) = \int_{\mathbb{R}^2} \frac{dx' dp'}{(4\pi)^2} \chi_{|x_0\rangle\langle x_0|}(x', p') e^{\frac{i}{2}x'p - \frac{i}{2}xp'} = \frac{\delta(x - x_0)}{4\pi}. \qquad (4.207)$$

These expressions lead us to the following characteristic and Wigner functions of the remaining modes:

$$\tilde{\chi}_{x_0}(\mathbf{r}_{\{N\}}) = \int_{\mathbb{R}^2} \frac{dp}{4\pi} \chi_{\hat{\rho}}(\mathbf{r}_{\{N\}}, 0, p) e^{-ipx_0/2}, \qquad (4.208)$$

$$\tilde{W}_{x_0}(\mathbf{r}_{\{N\}}) = \int_{\mathbb{R}} dp\, W_{\hat{\rho}}(\mathbf{r}_{\{N\}}, x_0, p). \qquad (4.209)$$

---

[12] As explained in detail after introducing axiom 6, for the measurement of a continuous observable this is an idealized situation taken here for simplicity.

Note that, even though the characteristic (4.206) and Wigner (4.207) functions of the projector are not normalizable (which makes sense, since the position eigenstate is also not), the characteristic and Wigner functions of the remaining modes can be normalized, so that the probability density function associated with the possible outcomes x_0 is given by

$$P(x_0) = \tilde{\chi}_{x_0}(\mathbf{0}) = \int_{\mathbb{R}^{2N}} \mathrm{d}^{2N} \mathbf{r}_{\{N\}} \tilde{W}_{x_0}(\mathbf{r}_{\{N\}}). \qquad (4.210)$$

Note that the projector $|x_0\rangle\langle x_0|$ is Gaussian, and therefore, partial homodyne measurements map Gaussian states into Gaussian states.

Heterodyne detection from homodyne detection

Along with photodetection (or on/off detection, which we will study in the next section), homodyne detection is one of the fundamental measurement schemes from which many other more complicated ones are generated. Among these, a most relevant example is *heterodyne detection*, described by a POVM[13] $\{\hat{\Pi}(\alpha) = |\alpha\rangle\langle\alpha|/\pi\}_{\alpha\in\mathbb{C}}$.

In loose terms, one could say that heterodyne detection corresponds to a measurement of the annihilation operator $\hat{a}$, since indeed it generates the suggestive probability distribution $P(\alpha) = \mathrm{tr}\{\hat{\Pi}(\alpha)\hat{\rho}\} = \langle\alpha|\hat{\rho}|\alpha\rangle$. However, note that the post-measurement state does not correspond in general to an eigenstate of $\hat{a}$.

Heterodyne detection can be implemented from homodyne detection as follows (see figure 4.8): the mode we want to measure is mixed on a 50/50 beam splitter with another mode in vacuum, and the two output modes are homodyned one in position $\hat{x}$ and the other in momentum $\hat{p}$. It is possible to show then that the probability of recording an outcome $\{x, p\}$ is precisely described by the POVM introduced above with $\alpha = (x + ip)/2$.

4.5.4 On/off detection and de-Gaussification by vacuum removal

Partial on/off detection

Despite the incredible advances in photodetector technologies, practical photon counters are still out of reach in most laboratories. In a very simplified manner, the problem arises from distinguishing when the photopulse, which is a macroscopic current created by random collisions between the free electrons and the ones bound to the metallic plates, was created from the amplification of n or $n + 1$ photo-electrons. Modern superconducting photodetectors, whose operating principle is not based on the photoelectric effect plus an amplification stage, allow for such photon-number resolution, but they are still far too expensive or come with other problems depending on what they are needed for in the particular experiment.

A less demanding detection strategy is the so-called on/off detection, in which the detector gives a signal whenever one or more photons reach it, but this signal is identical no matter how many photons triggered it. In the next section we will see

---

[13] It is immediate to show that the POVM is complete, that is, $\int_{\mathbb{C}} \mathrm{d}^2\alpha\hat{\Pi}(\alpha) = \hat{I}$, by expanding the coherent states in the Fock basis and integrating in polar complex coordinates $\alpha = r\exp(i\theta)$, i.e. $\int_{\mathbb{C}} \mathrm{d}^2\alpha = \int_0^\infty r\mathrm{d}r \int_0^{2\pi} \mathrm{d}\theta$.

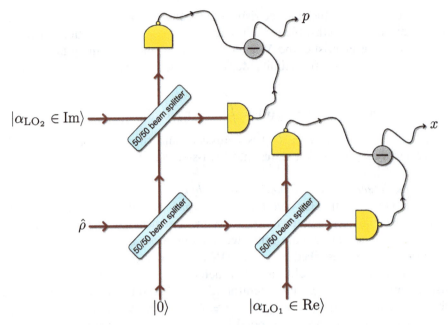

Figure 4.8. Heterodyne detection with a combination of two homodyne detection schemes.

that on/off detection allows for the implementation of interesting operations such as photon addition and subtraction onto a light field, which are the most fundamental non-Gaussian and non-unitary operations that one can think of.

On/off detection is then a projective measurement with two possible outcomes 'off' (no-click) and 'on' (click!), with the corresponding projectors

$$\hat{P}_{\text{off}} = |0\rangle\langle 0| \qquad \text{and} \qquad \hat{P}_{\text{on}} = \hat{I} - |0\rangle\langle 0| = \sum_{n=1}^{\infty} |n\rangle\langle n|. \qquad (4.211)$$

These projectors have very simple characteristic functions

$$\chi_{\hat{P}_{\text{off}}}(\mathbf{r}) = \chi_{|0\rangle\langle 0|}(\mathbf{r}) \qquad \text{and} \qquad \chi_{\hat{P}_{\text{on}}}(\mathbf{r}) = 4\pi\delta^{(2)}(\mathbf{r}) - \chi_{|0\rangle\langle 0|}(\mathbf{r}), \qquad (4.212)$$

and very simple Wigner functions as well,

$$W_{\hat{P}_{\text{off}}}(\mathbf{r}) = W_{|0\rangle\langle 0|}(\mathbf{r}) \qquad \text{and} \qquad W_{\hat{P}_{\text{on}}}(\mathbf{r}) = \frac{1}{4\pi} - W_{|0\rangle\langle 0|}(\mathbf{r}). \qquad (4.213)$$

Applied as a partial measurement (as usual, the measurement is applied onto the last mode of a system with $N + 1$ modes), on/off detection makes the system evolve from some initial state $\hat{\rho}$ of the $N + 1$ oscillators, to a reduced state of the first N oscillators with (unnormalized) characteristic and Wigner functions

$$\tilde{\chi}_{\text{off}}(\mathbf{r}_{\{N\}}) = \int_{\mathbb{R}^2} \frac{\mathrm{d}^2\mathbf{r}_{N+1}}{4\pi} \chi_{\hat{\rho}}(\mathbf{r}) \chi_{|0\rangle\langle 0|}(-\mathbf{r}_{N+1}), \qquad (4.214a)$$

$$\tilde{W}_{\text{off}}(\mathbf{r}_{\{N\}}) = 4\pi \int_{\mathbb{R}^2} \mathrm{d}^2\mathbf{r}_{N+1} W_{\hat{\rho}}(\mathbf{r}) W_{|0\rangle\langle 0|}(\mathbf{r}_{N+1}) \qquad (4.214b)$$

when the outcome is 'off' or

$$\tilde{\chi}_{\text{on}}(\mathbf{r}_{\{N\}}) = \chi_{\hat{\rho}_{\{N\}}}(\mathbf{r}_{\{N\}}) - \tilde{\chi}_{\text{off}}(\mathbf{r}_{\{N\}}), \tag{4.215a}$$

$$\tilde{W}_{\text{on}}(\mathbf{r}_{\{N\}}) = W_{\hat{\rho}_{\{N\}}}(\mathbf{r}) - \tilde{W}_{\text{off}}(\mathbf{r}_{\{N\}}), \tag{4.215b}$$

when the outcome is 'on', where $\hat{\rho}_{\{N\}} = \text{tr}_{N+1}\{\hat{\rho}\}$ is the reduced initial state of the non-measured modes.

De-Gaussification by vacuum removal
Note that $W_{\hat{\rho}_{\text{off}}}(\mathbf{r})$ is Gaussian, and therefore, the 'off' event projects the state of the non-measured modes into another Gaussian state. This is not the case for the 'on' event, whose associated Wigner function $W_{\hat{\rho}_{\text{on}}}(\mathbf{r})$ is not Gaussian, and therefore, it can be used as a *de-Gaussifying* operation. Moreover, it is a very convenient non-Gaussian operation from the theoretical point of view, because it is a sum of a constant term and a Gaussian, and hence it is still very easy to treat by extending the Gaussian formalism minimally.

In order to understand this better, let us analyze the case in which the initial state is a general Gaussian state $\hat{\rho}_{\text{G}}(\mathbf{d}, V)$ of the form (4.68) with $M = N$ and $M' = 1$, whose mean vector and covariance matrix we write as

$$\bar{\mathbf{r}} = (\bar{\mathbf{r}}_{\{N\}}, \bar{\mathbf{r}}_{N+1}) \qquad \text{and} \qquad V = \begin{bmatrix} V_{\{N\}} & C \\ C^T & V_{N+1} \end{bmatrix}, \tag{4.216}$$

where $\bar{\mathbf{r}}_{\{N\}} \in \mathbb{R}^{2N}$, $\bar{\mathbf{r}}_{N+1} \in \mathbb{R}^2$, $V_{\{N\}}$ and V_{N+1} are real, symmetric matrices of dimensions $2N \times 2N$ and 2×2, respectively, while C is a real $2N \times 2$ matrix. Using the Gaussian integral (4.59), it is straightforward to prove that the probability of the 'off' event is

$$p_{\text{off}} = \tilde{\chi}_{\text{off}}(\mathbf{0}) = \frac{2}{\sqrt{\det(V_{N+1} + I_{2\times2})}} \exp\left[-\frac{1}{8}\bar{\mathbf{r}}_{N+1}^T(V_{N+1} + I_{2\times2})\bar{\mathbf{r}}_{N+1}\right], \tag{4.217}$$

while the corresponding output state is the Gaussian $\hat{\rho}_{\text{off}} = \hat{\rho}_{\text{G}}(\bar{\mathbf{r}}_{\text{off}}, V_{\text{off}})$ with

$$\bar{\mathbf{r}}_{\text{off}} = \bar{\mathbf{r}}_{\{N\}} - C(V_{N+1} + I_{2\times2})^{-1}\bar{\mathbf{r}}_{N+1}, \tag{4.218a}$$

$$V_{\text{off}} = V_{\{N\}} + C(V_{N+1} + I_{2\times2})^{-1}C^T. \tag{4.218b}$$

The probability of the 'on' event is then $p_{\text{on}} = 1 - p_{\text{off}}$, which, based on (4.215), has an associated post-measurement Wigner function

$$W_{\text{on}}(\mathbf{r}_{\{N\}}) = \left(1 - p_{\text{off}}\right)^{-1}\left[W_{\hat{\rho}_{\text{G}}(\mathbf{d}_{\{N\}}, V_{\{N\}})}(\mathbf{r}_{\{N\}}) - p_{\text{off}} W_{\hat{\rho}_{\text{G}}(\mathbf{d}_{\text{off}}, V_{\text{off}})}(\mathbf{r}_{\{N\}})\right]. \tag{4.219}$$

Note that even though this Wigner function is not Gaussian, it is a simple combination of two Gaussians (specifically a 'negative mixture') and, hence, as already explained, this approach to de-Gaussification is very convenient since all the tools of Gaussian states and operations can be used.

Let us consider a simple example: we have two modes in the two-mode squeezed vacuum state (4.113), and we perform an on/off detection onto the second mode.

Given the photon-number correlation between the modes, it is obvious that whenever the outcome is 'off', the first mode is projected into the vacuum state, $\hat{\rho}_{\text{off}} = |0\rangle\langle 0|$. More interestingly, if the outcome is 'on' the state of the first mode will collapse to the mixture

$$\hat{\rho}_{\text{on}} \propto \sum_{n=1}^{\infty} \tanh^{2n} r |n\rangle\langle n|, \tag{4.220}$$

which is a thermal state with the vacuum component removed.

According to (4.217), and using the mean vector and covariance matrix of the two-mode squeezed vacuum state (4.114), the probability for the 'off' event is $P_{\text{off}} = 1/\cosh^2 r$, and it is simple to check that (4.219) leads to $\bar{\mathbf{r}}_{\text{off}} = \mathbf{0}$ and $V_{\text{off}} = I_{2\times2}$, corresponding to the vacuum state. The probability of the 'on' event reads then $P_{\text{on}} = \tanh^2 r$, and the corresponding Wigner function is

$$W_{\text{on}}(\mathbf{r}_1) = \sinh^{-2} r \left[\cosh^2 r \, W_{\hat{\rho}_{\text{th}}(\sinh^2 r)}(\mathbf{r}_1) - W_{|0\rangle\langle 0|}(\mathbf{r}_1) \right], \tag{4.221}$$

that is, a 'negative mixture' of a thermal and a vacuum state. It is simple to check that, at the origin of phase space, $\mathbf{r}_1 = \mathbf{0}$, the weight of the vacuum state is always larger than that of the thermal state, and hence this Wigner function always has a negative central region, surrounded by a positive one, which shows the non-Gaussian character of the state. In particular, we find that $W_{\text{on}}(\mathbf{0}) = -1/4\pi \cosh(2r)$. Moreover, this central negative region has more or less the same size irrespective of the squeezing value. Indeed, it is very simple to show that the radius of the central negative region is given by

$$\sqrt{\frac{2 \log_e(1 + \tanh^2 r)}{\tanh^2 r}}, \tag{4.222}$$

which is a monotonically increasing function of the squeezing, but is bounded by 1 from below and by $\log_e^{1/2} 4 \approx 1.18$ from above, so that it varies very little with r. In contrast, the positive region becomes larger as the squeezing increases, which arises from the thermal component of the state.

Finally, note that for a small squeezing parameter, the state tends to the $|1\rangle$ Fock state, as easily seen from (4.220). This is not an 'accident'; in fact, as we show in the next section, when a mode interacts very weakly via the two-mode squeezing or beam-splitter operations with a second mode in a vacuum, the 'on' detection of this second mode signals, respectively, the approximate application of the $\hat{a}^\dagger$ or $\hat{a}$ operators on the principal mode.

4.6 Non-Gaussian scenarios: photon addition, subtraction, and majorization properties of two-mode squeezed states

4.6.1 Photon addition and subtraction

In spite of their non-unitary character, photon addition and subtraction, which correspond to the application of $\hat{a}$ and $\hat{a}^\dagger$ as evolution onto a continuous-variable state, can be performed on a system conditioned to particular outcomes in a measurement,

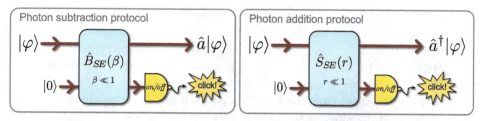

Figure 4.9. Protocols for the probabilistic implementation of photon subtraction and addition based on a click in an on/off detection.

that is, probabilistically. In this section we are going to explain how this can be achieved by using schemes based on a click in an on/off detection. Being able to apply these non-Gaussian, non-unitary operations onto a system opens the way to an exciting world in which many fundamental tests of quantum mechanics can be performed.

Consider a single mode S in some pure state $|\varphi\rangle$ (everything that follows is trivially generalized to general mixed states, or even to multi-mode states). As shown in figure 4.9, the subtraction protocol starts by mixing this mode via a weak beam-splitter transformation $\hat{B}_{SE}(\beta \ll 1)$ with another oscillator E in a vacuum state. We can make a series expansion of the beam-splitter operator (4.115a) up to first order in β, so that the joint state $|\varphi\rangle \otimes |0\rangle$ evolves into

$$|\psi\rangle_{SE} = |\varphi\rangle \otimes |0\rangle + \beta(\hat{a}|\varphi\rangle \otimes |1\rangle) + O(\beta^2), \qquad (4.223)$$

after the beam splitter. The last step of the protocol consists in applying an on/off measurement to the E mode, so that whenever the detector *clicks*, the state of the S oscillator collapses to the unnormalized state

$$\tilde{\rho} = \mathrm{tr}_E\left\{\left(\hat{I} \otimes \hat{P}_{\mathrm{on}}\right)|\psi\rangle_{SE}\langle\psi|\right\} = \beta^2 \hat{a}|\varphi\rangle\langle\varphi|\hat{a}^\dagger + O(\beta^4), \qquad (4.224)$$

which converges to the subtracted state $\hat{a}|\varphi\rangle$ as β goes to zero. Note, however, that the probability $p_{\mathrm{on}} = \mathrm{tr}\{\tilde{\rho}\}$ of the 'on' event vanishes in the $\beta \to 0$ limit, and therefore, one has to find a balance between having a non-zero *success probability* and a good approximation of the subtracted state. We will return to this point shortly.

The scheme that implements photon addition is very similar, except for the fact that now the beam-splitter transformation is replaced by a two-mode squeezing transformation $\hat{S}_{SE}(r \ll 1)$, see (4.109). For a small squeezing parameter, the operator can be expanded again to first order in the squeezing parameter r, leading to the joint state after the two-mode squeezer

$$|\psi\rangle_{SE} = |\varphi\rangle \otimes |0\rangle + r(\hat{a}^\dagger|\varphi\rangle \otimes |1\rangle) + O(r^2). \qquad (4.225)$$

Then, after a click in the on/off detector the unnormalized state of the S oscillator collapses to

$$\tilde{\rho} = \mathrm{tr}_E\left\{\left(\hat{I} \otimes \hat{P}_{\mathrm{on}}\right)|\psi\rangle_{SE}\langle\psi|\right\} = r^2 \hat{a}^\dagger|\varphi\rangle\langle\varphi|\hat{a} + O(r^4), \qquad (4.226)$$

which again converges to the added state $\hat{a}^\dagger|\varphi\rangle$ as r goes to zero.

Note that in the language of quantum operations, based on expressions (4.224) and (4.226), one would be tempted to define quantum operations $\mathcal{E}_{sub}$ and $\mathcal{E}_{add}$ associated with photon subtraction and addition, respectively, with corresponding Kraus operators $\hat{E}_{sub} = \beta\hat{a}$ and $\hat{E}_{add} = r\hat{a}^\dagger$. Note, however, that these Kraus operators do not lead to trace-decreasing quantum operations in general; they do so only provided that β and r are small enough. Indeed, note that denoting by $\bar{N} = \text{trace}\{\hat{a}^\dagger\hat{a}\hat{\rho}\}$ the mean photon number of the state, we obtain the success probabilities

$$p_{sub} = \text{tr}\left\{\hat{E}_{sub}^\dagger \hat{E}_{sub}\hat{\rho}\right\} = \beta^2\bar{N} \tag{4.227a}$$

$$p_{add} = \text{tr}\left\{\hat{E}_{add}^\dagger \hat{E}_{add}\hat{\rho}\right\} = r^2(\bar{N}+1). \tag{4.227b}$$

This shows that the operations are trace-decreasing, and hence physical, only when $\beta^2 < \bar{N}^{-1}$ and $r^2 < (\bar{N}+1)^{-1}$. Nevertheless, a more careful analysis of when the higher orders in the states (4.224) and (4.226) can be neglected would reveal that the conditions $\beta^2 \ll \bar{N}^{-1}$ and $r^2 \ll (\bar{N}+1)^{-1}$, and not just $\beta \ll 1$ and $r \ll 1$, are needed, so that the quantum operations $\mathcal{E}_{sub}$ and $\mathcal{E}_{add}$ are indeed trace-decreasing whenever they are accurately implemented. Note that this also means that the success probability has to be small, as we already pointed out above.

This simplified picture captures all the qualitative features that one needs to know about how the addition and subtraction operations can be applied probabilistically, that is, as trace-decreasing operations. Nevertheless, using the tools that we developed above, in particular the phase-space description of on/off detectors, one can treat them carefully from a quantitative point of view when benchmarking particular setups.

4.6.2 Increasing entanglement by local addition or subtraction

Throughout this book, we have introduced the idea that the entanglement of a state cannot be increased by acting locally on its entangled parts. However, this is only true deterministically, that is, as counterintuitive as it might seem, local strategies based on measurements can indeed enhance entanglement for some of the measurement outcomes (but, of course, on average the entanglement cannot increase). As an example of this, now we study how photon addition and subtraction can indeed achieve this task. In particular, we will study how the entanglement of the two-mode squeezed vacuum state increases when photon addition or subtraction are applied to either of the modes [31], which we denote by A and B, for Alice and Bob. It will be convenient to define the parameter $\lambda = \tanh r$.

Before addressing the problem, it will be useful to prove the following properties of the two-mode squeezed vacuum state under the action of annihilation and creation operators:

$$\hat{a}_B^k |r\rangle_{2sq} = \tanh^k(r)\hat{a}_A^{\dagger k} |r\rangle_{2sq}, \tag{4.228a}$$

$$\hat{a}_A^{\dagger k} |r\rangle_{2sq} = \cosh^k(r)\hat{S}_{AB}(r)\hat{a}_A^{\dagger k}|0\rangle. \tag{4.228b}$$

The first identity tells us that subtracting excitations in one mode of the pair is equivalent to adding excitations in the other when they are in a two-mode squeezed vacuum state. The second means that adding the excitations before or after the two-mode squeezing transformation is all the same. Let us now prove these relations.

The first one is easily proved for $k = 1$ by simply operating on the two-mode squeezed vacuum state (4.113) expanded in the Fock basis,

$$\hat{a}_B \,|r\rangle_{2sq} = \frac{1}{\cosh r}\sum_{n=1}^{\infty}\sqrt{n}\,\tanh^n r|n, n-1\rangle$$

$$= \frac{1}{\cosh r}\sum_{n=0}^{\infty}\sqrt{n+1}\,\tanh^{n+1} r|n+1, n\rangle = \tanh(r)\hat{a}_A^\dagger\,|r\rangle_{2sq}. \quad (4.229)$$

Since $\hat{a}_B$ and $\hat{a}_A^\dagger$ commute, iterating this identity we prove (4.228a) for any k. The second identity (4.228b) is also easily proven as follows:

$$\hat{S}_{AB}(r)\hat{a}_A^{\dagger k}|0, 0\rangle = \left[\hat{S}_{AB}^\dagger(-r)\hat{a}_A^\dagger \hat{S}_{AB}(-r)\right]^k \hat{S}_{AB}(r)|0, 0\rangle \underset{(4.110)}{\equiv} \left(\hat{a}_A^\dagger \cosh r - \hat{a}_B \sinh r\right)^k |r\rangle_{2sq}$$

$$\underset{\substack{\text{binomial}\\\text{expansion}}}{\equiv} \sum_{j=0}^{k}(-1)^{k-j}\binom{k}{j}\sinh^{k-j} r \cosh^j r\hat{a}_A^{\dagger j}\hat{a}_B^{k-j}\,|r\rangle_{2sq}$$

$$\underset{(4.228a)}{\equiv} \frac{1}{\cosh^k r}\sum_{j=0}^{k}(-1)^{k-j}\binom{k}{j}\sinh^{2k-2j} r \cosh^{2j} r\hat{a}_A^{\dagger k}\,|r\rangle_{2sq}$$

$$\underset{\substack{\text{binomial}\\\text{expansion}}}{\equiv} \frac{1}{\cosh^k r}\hat{a}_A^{\dagger k}\,|r\rangle_{2sq}, \quad (4.230)$$

where we have used the binomial expansion $(\hat{A} + \hat{B})^k = \sum_{j=0}^{k}\binom{k}{j}\hat{A}^j\hat{B}^{k-j}$, valid for any two commuting operators $\hat{A}$ and $\hat{B}$.

Equipped with these identities, we can now consider how the entanglement of the two-mode squeezed vacuum state is affected by photon addition and subtraction. To this aim, we start by defining the two-mode squeezed Fock state with k excitations in the first mode, which using the identities above is equivalent to a two-mode squeezed vacuum state with k-additions or subtractions in the first or second modes, respectively; explicitly:

$$\left|\Psi_\lambda^{(k)}\right\rangle = \hat{S}_{AB}(r)|k, 0\rangle = \frac{1}{\sqrt{k!}\,\cosh^k r}\hat{a}_A^{\dagger k}\,|r\rangle_{2sq} = \frac{1}{\sqrt{k!}\,\sinh^k r}\hat{a}_B^k\,|r\rangle_{2sq}. \quad (4.231)$$

In order to proceed, we write the state in the Fock basis using (4.113) and the second form of the state presented above. It is straightforward to obtain

$$\left|\Psi_\lambda^{(k)}\right\rangle = \sum_{n=0}^{\infty}\sqrt{p_n^{(k)}(\lambda)}\,|n+k, n\rangle, \quad (4.232)$$

with

$$p_n^{(k)}(\lambda) = (1 - \lambda^2)^{k+1} \lambda^{2n} \binom{n+k}{n}. \tag{4.233}$$

This is a very convenient representation of the state, since it is already in the Schmidt form. Therefore, its entanglement entropy can be evaluated as

$$E\left[\left|\Psi_\lambda^{(k)}\right\rangle\right] = -\sum_{n=0}^{\infty} p_n^{(k)} \log p_n^{(k)}. \tag{4.234}$$

In order to prove that this is a monotonically increasing function of k, we proceed as follows. Using the Pascal identity

$$\binom{n+k+1}{k+1} = \binom{n+k}{k+1} + \binom{n+k}{k}, \tag{4.235}$$

we can write

$$p_n^{(k+1)}(\lambda) = \lambda^2 p_{n-1}^{(k+1)}(\lambda) + (1 - \lambda^2) p_n^{(k)}(\lambda), \tag{4.236}$$

where we set $p_n^{(k)} = 0$ for $n < 0$ by definiteness. Next we use the strict concavity[14] of the function $h(x) = -x \log x$ to write

$$\sum_{n=0}^{\infty} h\left[p_n^{(k+1)}(\lambda)\right] > \lambda^2 \sum_{n=0}^{\infty} h\left[p_{n-1}^{(k+1)}(\lambda)\right] + (1 - \lambda^2) \sum_{n=0}^{\infty} h\left[p_n^{(k)}(\lambda)\right], \tag{4.237}$$

for $0 < \lambda < 1$. The final step consists in noting that $p_{n-1}^{(k+1)}$ is equivalent to $p_n^{(k+1)}$ up to a shift to the right in the Fock basis, which does not change the entropy, i.e. $\sum_{n=0}^{\infty} h[p_{n-1}^{(k+1)}] = \sum_{n=0}^{\infty} h[p_n^{(k+1)}]$, so that the previous expression is simply equivalent to

$$E\left[\left|\Psi_\lambda^{(k+1)}\right\rangle\right] > E\left[\left|\Psi_\lambda^{(k)}\right\rangle\right] \tag{4.238}$$

for $0 < \lambda < 1$. This shows that the entanglement of the two-mode squeezed vacuum state is enhanced by photon addition or subtraction, at least when acting on one mode only. Results for the simultaneous action on both modes can be consulted in the original reference [31].

4.6.3 Majorization properties of two-mode squeezed Fock states

As we discussed in section 3.4, we can use majorization theory to find relations between pure bipartite states which are stronger than simply 'which one is more entangled'. In particular, we saw that given two bipartite states $|\psi\rangle_{AB}$ and $|\varphi\rangle_{AB}$, $|\psi\rangle_{AB} \prec |\varphi\rangle_{AB}$ implies not only that $|\psi\rangle_{AB}$ is more entangled than $|\varphi\rangle_{AB}$, but also that it can be transformed into the latter deterministically via LOCC protocols. As an example of this in

---

[14] A real function is $f(x)$ is strictly concave when $f(\sum_{n=1}^{N} p_n x_N) > \sum_{n=1}^{N} p_n f(x_n)$ for any non-trivial probability distribution $\{p_n\}_{n=1,2,...,N}$ and set of non-equal points $\{x_n\}_{n=1,2,...,N}$.

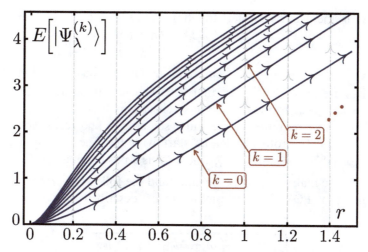

Figure 4.10. Entanglement entropy of the two-mode squeezed Fock states as a function of the squeezing parameter. The majorization relations are denoted in the curves, which imply stronger partial order relations between the states than the value of their entanglement (in particular, states with larger entanglement can be transformed into states with a lower one via deterministic LOCC protocols).

continuous-variable systems, we will now study the different majorization relations that exist between the two-mode squeezed Fock states defined above $|\Psi_\lambda^{(k)}\rangle$.

We are going to discuss two types of majorization relations [32], which are summarized in figure 4.10:

$$|\Psi_\lambda^{(k)}\rangle \prec |\Psi_\lambda^{(k' < k)}\rangle \qquad \text{and} \qquad |\Psi_\lambda^{(k)}\rangle \prec |\Psi_{\lambda' < \lambda}^{(k)}\rangle. \qquad (4.239)$$

The first one shows that for a fixed squeezing λ, a two-mode squeezed k-Fock state can be transformed via an LOCC protocol into states with smaller k. Similarly, the second one means that a two-mode squeezed Fock state can be transformed into states with smaller squeezing parameter.

This result can be proven in two ways: either by finding explicit LOCC protocols performing the corresponding transformation between the states, or the column-stochastic matrices connecting the corresponding Schmidt distributions. Regarding the column-stochastic matrices, the following results can be found [32]:

$$\mathbf{p}^{(k)}(\lambda) = D^{(k-k')}(\lambda)\mathbf{p}^{(k')}(\lambda), \qquad (4.240a)$$

$$\mathbf{p}^{(k)}(\lambda) = R^{(k)}(\lambda, \lambda')\mathbf{p}^{(k)}(\lambda'), \qquad (4.240b)$$

with matrices D and R having matrix elements

$$D_{nm}^{(k-k')}(\lambda) = (1 - \lambda^2)^{k-k'} \binom{m + k - k' - 1}{k - k' - 1} \lambda^{2(n-m)} H(n - m), \qquad (4.241a)$$

$$R_{nm}^{(k)}(\lambda, \lambda') = \binom{m + k}{m}^{-1} \left(\frac{1 - \lambda^2}{1 - \lambda'^2}\right) \left[L_{n-m}^{(k,m)}\lambda^2 - L_{n-m-1}^{(k,m+1)}\lambda'^2\right] \lambda'^{2(n-m-1)} H(n - m), \quad (4.241b)$$

where $H(x)$ is the Heaviside step function defined as $H(x) = 0$ for $x < 0$ and $H(x) = 1$ for $x \geqslant 0$, and $L_m^{(k,n)} = n\binom{n+k}{k}\binom{m+k}{k}\lambda'^{-2n}B(\lambda'^2; n, 1+k)$,

$B(z; a, b) = \int_0^z \mathrm{d}x\, x^{a-1}(1-x)^{b-1}$ being the incomplete beta function. Note that both matrices are lower-triangular, that is, their elements above the diagonal are zero. This is a typical form of column-stochastic matrices in infinite dimension, which makes them a bit more simple and systematic to find recursively.

As for the LOCC protocols, let us start with the one corresponding to the transformation $|\Psi_\lambda^{(k+\Delta k)}\rangle \xrightarrow[\text{LOCC}]{} |\Psi_\lambda^{(k)}\rangle$, where Δk is a positive integer. Inspired by the column-stochastic matrix (4.240a), we can build the following protocol. Bob starts by performing a POVM-based measurement described by the measurement operators

$$\hat{B}_m = \sum_{l=m}^\infty \sqrt{\frac{(1-\lambda^2)^{\Delta k}\binom{m+\Delta k-1}{\Delta k-1}\lambda^{2m}p_{l-m}^{(k)}}{p_l^{(k+\Delta k)}}}\,|l-m\rangle\langle l|. \qquad (4.242)$$

It is easy to verify the condition $\sum_{m=0}^\infty \hat{B}_m^\dagger \hat{B}_m = \hat{I}$ by using the (4.240a). Suppose that Bob obtains the outcome m, after which the state collapses to

$$\left(\hat{I} \otimes \hat{B}_m\right)|\Psi_\lambda^{(k+\Delta k)}\rangle \propto \sum_{l=m}^\infty \sqrt{p_{l-m}^{(k)}}|l+k+\Delta k, l+m\rangle = \sum_{n=0}^\infty \sqrt{p_n^{(k)}}|n+k+m+\Delta k, n\rangle.$$

$$(4.243)$$

Bob then communicates the outcome m of his measurement to Alice, who performs the shift operation

$$\hat{A}_m = \sum_{l=0}^\infty |l\rangle\langle l+m+\Delta k|, \qquad (4.244)$$

yielding the desired state $|\Psi_\lambda^{(k)}\rangle$ for all m. Note that the shift operator is trace preserving in the subspace spanned by $\{|j+m+\Delta k\rangle\}_{j=0,1,..}$, which is the support of $(\hat{I} \otimes \hat{B}_m)|\Psi_\lambda^{(k+\Delta k)}\rangle$ on Alice's space.

Let us move on now to the LOCC protocol associated to the transformation $|\Psi_\lambda^{(k)}\rangle \xrightarrow[\text{LOCC}]{} |\Psi_{\lambda'<\lambda}^{(k)}\rangle$. Given the complicated form of the column-stochastic matrix $R^{(k)}(\lambda, \lambda')$, it seems hard in this case to get inspiration from the relation (4.240b). However, in [32] a successful LOCC strategy was found, which is the one we will discuss now. The protocol starts with Bob mixing his mode B with an ancillary mode C on a beam splitter of transmissivity T. The initial state is written in the Fock basis as

$$|\psi\rangle_{ABC} = |\Psi_\lambda^{(k)}\rangle \otimes |0\rangle = \mathcal{N}(k,\lambda)\sum_{n=0}^\infty \lambda^n \binom{n+k}{k}^{1/2}|n+k, n, 0\rangle, \qquad (4.245)$$

with $\mathcal{N}(k, \lambda) = (1 - \lambda^2)^{(k+1)/2}$, which becomes

$$|\psi'\rangle_{ABC} = \mathcal{N}(k, \lambda) \sum_{n,m=0}^{\infty} \sqrt{(T\lambda^2)^n \left(\frac{1-T}{T}\right)^m \binom{n+k}{k}\binom{n}{m}} |n+k, n-m, m\rangle, \quad (4.246)$$

after passing through the beam splitter. The second step consists on Bob measuring the number of photons reflected by the beam splitter, that is, on mode C. With probability

$$\mathcal{P}(l) = {}_{ABC}\langle\psi'| \left(\hat{I} \otimes \hat{I} \otimes |l\rangle\langle l|\right) |\psi'\rangle_{ABC} = (1 - T)^l \lambda^{2l} \binom{k+l}{l} \frac{\mathcal{N}^2(k, \lambda)}{\mathcal{N}^2\left(k+l, \sqrt{T}\lambda\right)},$$

$$(4.247)$$

Bob will obtain l photons as an outcome, and the state of modes A and B will collapse in that case to

$$\sqrt{\mathcal{P}(l)}\, |\psi''\rangle_{AB} = {}_C\langle l|\psi'\rangle_{ABC}$$

$$= \mathcal{N}(k, \lambda)\left(\frac{1-T}{T}\right)^{l/2} \sum_{n=l}^{\infty} (T\lambda^2)^{n/2}\binom{n+k}{k}^{1/2}\binom{n}{l}^{1/2} |n+k, n-l\rangle. \quad (4.248)$$

Making the variable change $n - l \to n$ in the sum, and using the relation

$$\binom{n+l+k}{k}\binom{n+l}{l} = \binom{n+k+l}{n}\binom{k+l}{l}, \quad (4.249)$$

this state can be rewritten as

$$\sqrt{\mathcal{P}(l)}\, |\psi''\rangle_{AB} = \mathcal{N}(k, \lambda)(1 - T)^{l/2}\lambda^l \binom{k+l}{l}^{1/2} \sum_{n=0}^{\infty} (T\lambda^2)^{n/2}\binom{n+k+l}{n}^{1/2} |n+k+l, n\rangle$$

$$= \sqrt{\mathcal{P}(l)}\left|\Psi_{\sqrt{T}\lambda}^{(k+l)}\right\rangle. \quad (4.250)$$

Therefore, by properly choosing the transmissivity of the beam splitter so that $\lambda' = \sqrt{T}\lambda$, the final state is $|\Psi_{\lambda'}^{(k+l)}\rangle$. The last step of the protocol consists then on the application of the LOCC-based transformation $|\Psi_{\lambda'}^{(k+l)}\rangle \to |\Psi_{\lambda'}^{(k)}\rangle$ that we introduced above. Note that the proposed protocol requires two steps,

$$|\Psi_\lambda^{(k)}\rangle \xrightarrow[\text{LOCC}]{\text{probabilistic}} |\Psi_{\lambda'}^{(k+l)}\rangle \quad \text{and then} \quad |\Psi_{\lambda'}^{(k+l)}\rangle \xrightarrow[\text{LOCC}]{\text{deterministic}} |\Psi_{\lambda'}^{(k)}\rangle, \quad (4.251)$$

meaning that it is not the minimal one, since according to what we saw in section 3.4.3, there always exists a one-way-direct LOCC protocol. This is an example of the clear dichotomy that exists in the world of quantum information (and mathematics in general): it is very useful to know that something exists, but that does not always mean that we know how to find it.

Bibliography

[1] Schleich W P 2001 *Quantum Optics in Phase Space* (New York: Wiley)
[2] Simon R, Mukunda N and Dutta B 1994 Quantum-noise matrix for multimode systems: $U(n)$ invariance, squeezing, and normal forms *Phys. Rev.* A **49** 1567

[3] Arvind Dutta B, Mukunda N and Simon R 1995 The real symplectic groups in quantum mechanics and optics *Pramana J. Phys.* **45** 471

[4] Navarrete-Benlloch C 2011 Contributions to the quantum optics of multimode optical parametric oscillators *PhD Thesis* arXiv:1504.05917

[5] Gerry C C and Knight P L 2005 *Introductory Quantum Optics* (Cambridge: Cambridge University Press)

[6] Braunstein S L 2005 Squeezing as an irreducible resource *Phys. Rev.* A **71** 055801

[7] Williamson J 1936 On the algebraic problem concerning the normal forms of linear dynamical systems *Am. J. Math.* **58** 141

[8] Serafini A, Illuminati F and De Siena S 2004 Symplectic invariants, entropic measures and correlations of Gaussian states *J. Phys.* B **37** L21

[9] Duan L-M, Giedke G, Cirac J I and Zoller P 2000 Inseparability criterion for continuous variable systems *Phys. Rev. Lett.* **84** 2722

[10] Einstein A, Podolsky B and Rosen N 1935 Can quantum-mechanical description of physical reality be considered complete? *Phys. Rev.* **47** 777

[11] Bell J S 1964 On the Einstein Podolsky Rosen paradox *Physics* **1** 195

[12] Bell J S 2004 *Speakable and Unspeakable in Quantum Mechanics: Collected Papers on Quantum Philosophy* (Cambridge: Cambridge University Press)

[13] Freedman S J and Clauser J F 1972 Experimental test of local hidden-variable theories *Phys. Rev. Lett.* **28** 938

[14] Aspect A, Dalibard J and Roger G 1982 Experimental test of Bells inequalities using time-varying analyzers *Phys. Rev. Lett.* **49** 1804

[15] Brunner N, Cavalcanti D, Pironio S, Scarani V and Wehner S 2014 Bell nonlocality *Rev. Mod. Phys.* **86** 419

[16] Hensen B *et al* 2015 Experimental loophole-free violation of a Bell inequality using entangled electron spins separated by 1.3 km *Nature* **526** 682

[17] Giustina M *et al* 2015 Significant-loophole-free test of Bell's theorem *Phys. Rev. Lett.* **115** 250401

[18] Shalm L K *et al* 2015 Strong loophole-free test of local realism *Phys. Rev. Lett.* **115** 250402

[19] Simon R 2000 Peres-Horodecki separability criterion for continuous variable systems *Phys. Rev. Lett.* **84** 2726

[20] Werner R F and Wolf M M 2001 Bound entangled Gaussian states *Phys. Rev. Lett.* **86** 3658

[21] Ferraro A, Olivares S and Paris M G A 2005 *Gaussian States in Quantum Information* (*Napoli Series on Physics and Astrophysics*) (Naples: Bibliopolis)

[22] Giedke G, Kraus B, Lewenstein M and Cirac J I 2001 Entanglement criteria for all bipartite Gaussian states *Phys. Rev. Lett.* **87** 167904

[23] Vidal G and Werner R F 2002 Computable measure of entanglement *Phys. Rev.* A **65** 032314

[24] Holevo A S and Werner R F 2001 Evaluating capacities of bosonic Gaussian channels *Phys. Rev.* A **63** 032312

[25] Caruso F, Eisert J, Giovannetti V and Holevo A S 2008 Multi-mode bosonic Gaussian channels *New J. Phys.* **10** 083030

[26] Caruso F, Eisert J, Giovannetti V and Holevo A S 2011 Optimal unitary dilation for bosonic Gaussian channels *Phys. Rev.* A **84** 022306

[27] Bernstein D S 2005 *Matrix Mathematics: Theory Facts, and Formulas* (Princeton, NJ: Princeton University Press)

[28] Mollow B R 1968 Quantum theory of field attenuation *Phys. Rev.* **168** 1896
[29] Louisell W H 1973 *Quantum Statistical Properties of Radiation* (New York: Wiley)
[30] Puri R R 2001 *Mathematical Methods of Quantum Optics* (Berlin: Springer)
[31] Navarrete-Benlloch C, García-Patrón R, Shapiro J H and Cerf N J 2012 Enhancing quantum entanglement by photon addition and subtraction *Phys. Rev.* A **86** 012328
[32] García-Patrón R, Navarrete-Benlloch C, Lloyd S, Shapiro J H and Cerf N J 2012 Majorization theory approach to the Gaussian channel minimum entropy conjecture *Phys. Rev. Lett.* **108** 110505